SpringerBriefs in Electrical and Computer Engineering

Signal Processing

Series Editors

Woon-Seng Gan, Singapore
C.-C. Jay Kuo, Los Angeles, USA

For further volumes:
http://www.springer.com/series/11560

Y.-W. Peter Hong · Pang-Chang Lan
C.-C. Jay Kuo

Signal Processing Approaches to Secure Physical Layer Communications in Multi-Antenna Wireless Systems

 Springer

Y.-W. Peter Hong
Department of Electrical Engineering
National Tsing Hua University
Hsinchu
Taiwan, R.O.C.

C.-C. Jay Kuo
Department of Electrical Engineering
University of Southern California
Los Angeles, CA
USA

Pang-Chang Lan
Department of Electrical Engineering
University of Southern California
Los Angeles, CA
USA

ISSN 2196-4076   ISSN 2196-4084   (electronic)
ISBN 978-981-4560-13-9   ISBN 978-981-4560-14-6   (eBook)
DOI 10.1007/978-981-4560-14-6
Springer Singapore Heidelberg New York Dordrecht London

Library of Congress Control Number: 2013944087

Printed on acid-free paper

Springer is part of Springer Science+Business Media (www.springer.com)

To my beloved family

—Peter

Dedicated to my parents

—Pang-Chang

In memory of my beloved father

—Jay

Preface

Physical layer secrecy has received much attention in recent years, especially in wireless communications, due to the rapidly increasing data traffic and high demand for ubiquitous connectivity. Different from conventional cryptographic approaches used to address wireless security issues, physical layer secrecy utilizes channel coding and signal processing techniques to communicate secret messages between the source and the destination while maintaining confidentiality against the eavesdropper. These studies originate from the information theory literature, where the focus is often to determine the existence of channel codes that can achieve this task or to derive the fundamental limit on the maximum code rate that can be applied reliably under the secrecy constraint (i.e., the secrecy capacity). In particular, information-theoretic results have shown that the secrecy capacity increases with the difference between the reception quality at the destination and that at the eavesdropper. Motivated by this result, signal processing approaches have been developed in both the data transmission and channel estimation phases to maximize the signal quality difference between the destination and the eaves-dropper. In general, both coding and signal processing aspects of the problem must be taken into consideration in order to achieve the secrecy capacity. However, when the optimal joint design is unknown, signal processing techniques can also be used to help increase the achievable secrecy rate or reduce the complexity of the channel coding operations. These approaches are especially interesting in multi-antenna wireless systems, where the spatial degrees of freedom can be exploited to further enhance secrecy.

This book provides an overview of signal processing approaches that can be used to enhance physical layer secrecy in multi-antenna wireless systems. Specifically, in the data transmission phase, we review secrecy beamforming and precoding techniques that can be used to not only strengthen the signal at the destination, but also reduce the information leakage to the eavesdropper. We also consider the use of artificial noise along with the transmission of the information signal to further degrade the reception quality at the eavesdropper. Moreover, we further extend the use of these techniques to distributed antenna or relay systems where the multiple antennas may not be located at a single terminal. The additional spatial degrees of freedom provide more design flexibility and performance gains, but may also increase the security threats due to additional transmission required for coordination among the distributed terminals and also due to the trustworthiness of system members. In terms of channel estimation, we review the so-called discriminatory channel estimation scheme which utilizes the design of artificial noise aided training signals to increase the difference between the effective signal quality at the destination and that at the eavesdropper. In this case, the desired signal quality discrepancy is enhanced even before the transmission of the confidential message and need only be done once every coherence interval (instead of every symbol period).

This book is divided into 6 chapters. Chapter 1 introduces the importance and challenge of achieving secrecy in wireless systems, and provides a brief mention of the background on different notions of physical lay secrecy. Chapter 2 briefly summarizes the basic results in information-theoretic secrecy. In Chapter 3, secrecy beamforming and precoding designs along with the use of artificial noise are described as ways to effectively enhance the desired signal quality discrepancy in the data transmission phase. In Chapter 4, these techniques are extended to distributed antenna or relay systems. In Chapter 5, discriminatory channel estimation schemes are described as methods to achieve signal quality discrepancy in the channel estimation phase. Finally, in Chapter 6, several applications of physical layer secrecy and their respective research directions are further introduced.

The purpose of this book is to highlight the role of signal processing in achieving secrecy in the physical layer, and to provide researchers and graduate students, who are interested in pursuing research in this area, an overview of the standard techniques known today and a basic understanding of the challenges that may arise in this area. We would like to remark that, due to the vast literature in this area, it is not possible to give a comprehensive treatment of the material in the literature. However, we hope that the materials included in this book provide a logical treatment of the basic results and allow researchers and graduate students to gain sufficient knowledge to pursue research in this area.

Hsinchu, Taiwan, May 2013   Y.-W. Peter Hong
Los Angeles, USA, May 2013   Pang-Chang Lan
                                                               C.-C. Jay Kuo

Acknowledgments

This work was supported in part by the National Science Council, Taiwan, under grant NSC 100-2628-E-007-025-MY3. The authors would also like to thank Ta-Yuan Liu, Yu-Ching Chen, Li-Ren Chen, and Lin-Ming Huang for extensive proof-reading of the manuscript.

Contents

Chapter 1
Introduction

Abstract This chapter provides a brief introduction of security issues that may arise in wireless communication systems and describe physical layer techniques that can be used to address these issues. Different notions of physical layer secrecy are introduced, including keyless transmission of confidential messages (which is the focus of this book), channel-based secret key generation, and signal transmissions with low probability of interception and detection. Backgrounds on these techniques as well as an overview of the book content are provided.

Keywords Security · Cryptography · Physical layer secrecy · Secret key generation · Low probability of interception (LPI) · Low probability of detection (LPD).

1.1 Security in Wireless Communication Systems

With the increasing demand for mobility and ubiquitous connectivity, wireless communications is playing an integral part in our daily lives and is having a significant impact on society. Confidential and private information, such as e-banking, e-commerce, and medical information, is part of the mass data being transmitted over the wireless medium. However, due to the broadcast nature of wireless transmissions, communication over the wireless medium is often vulnerable to signal interception or eavesdropping by unauthorized receivers, as depicted in Fig. 1.1. Security and privacy issues have thus drawn much attention from both industry and academia, but many problems remain to be open and challenging.

In general, security in wireless networks may involve many tasks, including confidentiality, authentication, integrity, access control, and availability, etc. [1, 2]. Confidentiality refers to the prevention of unauthorized disclosure of information; authentication refers to the confirmation of the identity of different terminals; integrity ensures that the transmitted information is not illegally modified; access control and availability prevent denial-of-service (DoS) attacks. Conventionally, these

Y.-W. P. Hong et al., *Signal Processing Approaches to Secure Physical Layer Communications in Multi-Antenna Wireless Systems*, SpringerBriefs in Signal Processing, DOI: 10.1007/978-981-4560-14-6_1, © The Author(s) 2014

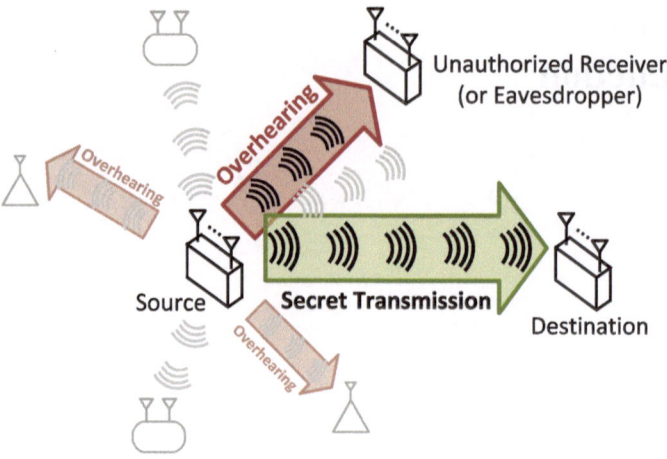

Fig. 1.1 The risk of overhearing in secret communications over the wireless medium

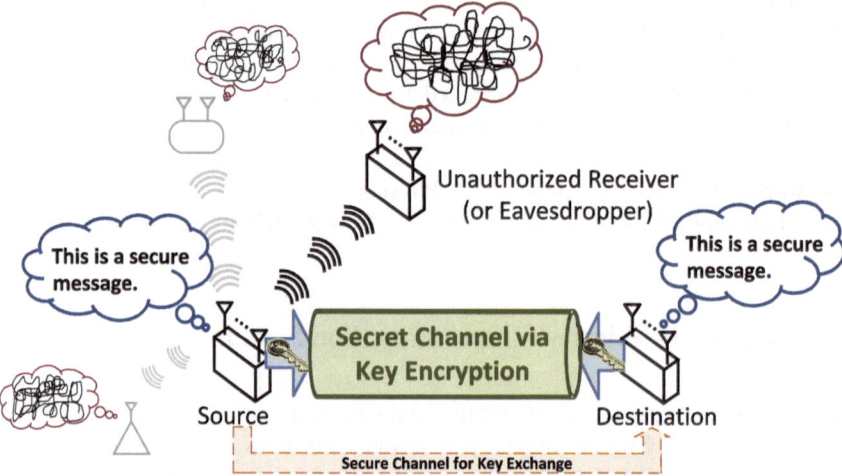

Fig. 1.2 Illustration of symmetric-key cryptography, which is used to construct a secrecy channel between the source and the destination for transmission of a confidential message, but requires a secure channel or protocol for the exchange of secret keys

issues have been addressed mostly in the upper layers of the network protocol stack using cryptographic encryption and decryption methods, e.g., data encryption standard (DES) [3] and advanced encryption standard (AES) [4]. When symmetric-key cryptosystems are employed, as illustrated in Fig. 1.2, a common private key is to be shared by two users and is used to encrypt and decrypt the confidential message. However, this requires a secure channel or protocol, such as the Diffie-Hellman [5] key exchange protocol, for the exchange of shared secret keys. The difficulties in secret key distribution and management [6] lead to security vulnerabilities in wireless systems. Alternatively, public-key cryptosystems, such as RSA [7], allow the use of a public key for encryption and a separate private key for decryption. The public key is available to all users whereas the private key is known only to the receiver. However, the security achieved by the cryptographic methods mentioned above relies on the computational hardness of decrypting the message when the secret key is not available. As the computational power increases, e.g., with the development of quantum computers, the computational hardness of certain mathematical problems, for which the encryption and decryption are based on, may no longer hold, causing many current cryptosystems to break down.

In recent years, many coding and signal processing techniques in the physical layer have been developed to support and to further enhance security in wireless systems. These techniques include, e.g., keyless physical layer secrecy transmission schemes [8–10], channel-based secret key generation schemes [11], and signal designs with low probability of interception and detection [12]. Different from conventional cryptographic methods, where the fast channel variations and the broadcast nature of the wireless medium may cause additional challenges to their design, these physical layer techniques exploit (rather than avoid) properties of wireless transmissions to better secure the communication channel. In particular, spatial variations of the channel are utilized to ensure that signals received at different locations are not the same; temporal variations of the channel are essential for destinations (that, on average, experience worse channel conditions than the eavesdropper) to temporarily experience better channel conditions at certain time instants; and the broadcast nature of wireless transmissions makes possible the emission of jamming signals to degrade the eavesdropper's reception. These physical layer techniques are used to support and complement security protocols in the upper layers of the network protocol stack, but are not meant as replacements for conventional cryptographic approaches. These physical layer techniques are mentioned in more detail in the following section.

It is worthwhile to note that, while many of the security issues mentioned above (e.g., authentication, integrity, and availability) are equally important, this book focuses on the issue of maintaining confidentiality of information transfer. Moreover, we consider only the case of passive adversaries, that only aim to intercept the confidential message or to detect the transmission activities but do not actively transmit signals. In cases with active adversaries, different attacks, such as jamming, impersonation, and message modification, can also be performed to limit the achievable secrecy. However, these topics are beyond the scope of this book. Readers are referred to [1, 2] for further discussions on these topics.

1.2 Background on Physical Layer Secrecy

In this section, we discuss briefly the three physical layer secrecy techniques mentioned in the previous section, namely, keyless physical layer secrecy transmission, channel-based secret key generation, and signal designs with low probability of interception and detection. Special emphasis is given to keyless physical layer secrecy transmissions as it is the focus of this book.

1.2.1 Keyless Physical Layer Secrecy Transmissions

Keyless physical layer secrecy transmission schemes were first studied in the context of wiretap channels by Wyner in [8], and was later extended to Gaussian channels in [9] and to broadcast channels with confidential messages in [10]. Here, confidential messages are transmitted using channel coding schemes (with random binning and channel prefixing techniques) to allow for reliable decoding at the destination while achieving substantial confusion at the eavesdropper. Early works in this field appeared mostly in the information theory literature and were mostly concerned with the so-called secrecy capacity, which is defined as the maximum achievable rate between the source and the destination while ensuring that no information can be inferred by the eavesdropper. It was shown that a non-zero secrecy capacity can be achieved between the source and the destination if the channel to the destination is better than that to the eavesdropper. These studies demonstrated the possibility of using properties of physical channels (without the use of secret keys) to ensure confidentiality of the transmitted messages against the eavesdropper. This avoids the inherent vulnerabilities caused by key distribution and management in conventional cryptographic systems.

Motivated by the emergence of wireless applications, keyless physical layer transmission schemes were also examined in wireless systems, where the dynamic nature of fading channels must also be taken into consideration [13, 14]. Specifically, it was shown in [13, 14] that, by exploiting the temporal variations of the channel, a positive secrecy rate can be achieved even when the average channel to the destination is of lower average quality than that to the eavesdropper. Extensions to multiple-input multiple-output (MIMO) wiretap channels, where the source, the destination, and the eavesdropper are all equipped with multiple antennas, were also examined recently in, e.g., [15–18]. The additional degrees of freedom provided by the multiple antennas in this case can be exploited to further enhance secrecy in the physical layer. In particular, as shown in [16, 18], this can be done by first employing secrecy precoding techniques to decompose the channel into multiple parallel sub-channels and by using wiretap codes over sub-channels that yield better quality to the destination than to the eavesdropper. However, this scheme requires perfect knowledge of the main and eavesdropper channels at the source, which may not be achievable in practice. When the eavesdropper's channel is unknown, artificial noise (AN) can also be emitted on top of the information-bearing signal to disrupt the reception at the eavesdropper, as illustrated in Fig. 1.3. With multiple antennas at the source, AN can be placed in dimensions that cause least interference at the destination. By doing so, the difference between the signal qualities at the destination and the eavesdropper (and, thus, the achievable secrecy rate) can be effectively increased. The desired spatial degrees of freedom can also be provided by relays or distributed antenna systems, and secrecy precoding and AN techniques can be applied accordingly.

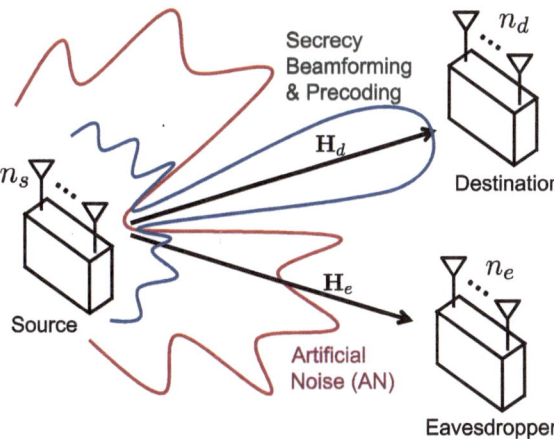

Fig. 1.3 Illustration of secrecy beamforming/precoding and artificial noise usage in a wireless multi-antenna system in the presence of an eavesdropper

1.2.2 Channel-Based Secret Key Generation

Channel-based secret key generation schemes [11] utilize unique characteristics of the channel between two users (i.e., the source and the destination) as the common randomness to generate symmetric keys locally at the two terminals. Here, channel knowledge is typically obtained by having the two terminals transmit training sequences to each other and by having the terminals perform channel estimation locally based on their received training signals. By assuming that the channel between the two terminals is reciprocal, the channel estimate at both ends will be approximately the same and can be utilized as the common random seed for secret key generation. However, these channel estimates are often subject to discrepancies caused by noise and may lead to key disagreement. Hence, key reconciliation and privacy amplification methods are necessary for error detection and correction. Since an eavesdropper located more than half a wavelength away will experience independent fading, it will not be able to infer the common secret key generated by the source and the destination.

The use of common randomness to generate secret keys at different terminals was first examined in [19, 20]. The use of channel characteristics as the common randomness was examined more recently, e.g., in [21–24]. These schemes utilize quantization of the amplitude and/or phase of the channel to mitigate the effect of noise and to determine the common index of the secret key at the two terminals. Similar concepts were also utilized to generate group keys in [25]. The performance of channel-based secret key generation schemes is often measured by the key generation rate, the key entropy, and the key disagreement probability. These three measures are often limited by properties of the physical channel, such as the channel coherence time and the channel quality. Techniques that can achieve the best tradeoff between secret key generation rate and key disagreement probability are important and have attracted many research efforts in recent years. Readers are referred to [11] for a survey on these techniques.

1.2.3 Signal Designs with Low Probability of Interception and Detection

Signal designs that can achieve low probability of interception (LPI) and low probability of detection (LPD) were the focus of many studies on security in the past in the signal processing literature, e.g., in [26, 27]. These tasks were often achieved using spread spectrum technology [28–30], with which the signal is spread over a frequency range that is much wider than its original bandwidth. These examples include the use of direct-sequence code division multiple access (DS-CDMA) [31], where signals are multiplied by pseudo-noise sequences and hidden in the background noise to reduce the probability of detection, and also the use of frequency-hopping (FH) [32], where the center frequency of the signal hops randomly over a wide frequency range to increase the difficulty of intercepting or jamming the transmission. FH is currently being used in many military as well as commercial applications, such as bluetooth or cordless phones.

More recently, the exploitation of spatial and temporal diversity to achieve LPI and LPD in multi-antenna wireless systems has been examined in [12]. By assuming that the destination has knowledge of its own channel, the work in [12] investigated the channel capacity and the capacity achieving transmission strategy between the source and the destination subject to constraints on the LPI and the LPD. The user-dependent channel information is viewed as a form of spatial encryption with the channel coefficients as the secret key. This "secret key" is distributed to the users via the transmission of training sequences in the channel estimation phase.

1.3 Book Overview

This book reviews various signal processing techniques used to support keyless physical layer secrecy transmissions in multi-antenna wireless systems. Motivated by results in information theory, which show that the secrecy capacity increases with the difference in quality between the source to destination and the source to eavesdropper channels, signal processing techniques in both the data transmission and the channel estimation phases have been developed in the literature to help construct effective channels at the destination and the eavesdropper with large channel quality discrepancies. These techniques are reviewed in this book.

In Chap. 2, we provide a brief review of the key results on information-theoretic secrecy. This includes studies of the secrecy capacity of different wiretap channels, e.g., the discrete memoryless wiretap channel [8, 10], the Gaussian wiretap channel [9], and the multi-antenna Gaussian wiretap channel [15–18], and the compound wiretap channel [33].

In Chap. 3, we focus on the data transmission phase and describe various secrecy beamforming and precoding schemes [15–18] that can be used to enlarge the difference in the signal qualities at the destination and the eavesdropper. We also discuss the use of artificial noise (AN) [15, 16, 34] on top of the beamformed or precoded signal to degrade the reception at the eavesdropper. AN is properly chosen to maintain quality performance at the destination. The generalization of these techniques to cases with multiple destinations and eavesdroppers is also discussed.

In Chap. 4, we extend the discussions on secrecy beamforming and precoding to relay and distributed antenna systems [35–38]. The use of relays provides additional spatial degrees of freedom that can be further exploited to enhance secrecy. The relays not only help to forward data to the destination but also can emit AN or jamming signals to disrupt the reception at the eavesdropper. However, with an additional node participating in the transmission, one must take into consideration the additional risk of information leakage due to the additional transmission required to communicate with the relay and also due to the trustworthiness of the relay.

In Chap. 5, we shift our focus to the channel estimation phase and describe training procedures that can be used to discriminate the channel estimation performance at the destination and the eavesdropper. By not allowing the eavesdropper to obtain a good estimate of its channel, its effective SNR would be poor and its ability to intercept the message in the data transmission phase would deteriorate. This concept was exploited in the so-called discriminatory channel estimation (DCE) scheme proposed in [39, 40]. Two DCE schemes are introduced, namely, the feedback and retraining and the two-way training DCE schemes.

In Chap. 6, we describe briefly the application of the physical layer secrecy techniques introduced in previous chapters to modern applications, such as cognitive radio, orthogonal frequency-division multiplexing (OFDM) systems, and ad hoc networks. Future research directions are also discussed.

References

1. Lou W, Ren K (2009) Security, privacy, and accountability in wireless access networks. IEEE Wirel Commun 16(4):80–87
2. Shiu Y S, Chang S-Y, Wu H-C, Huang SC-H, Chen H-H (2011) Physical layer security in wireless networks: a tutorial. IEEE Wirel Commun 18(2):66–74
3. Data Encryption Standard FIPS-46, National Bureau of Standards Std., Jan 1977
4. Advanced Encryption Standard FIPS-197, National Bureau of Standards and Technology Std., Nov 2001
5. Diffie W, Hellman ME (1976) New directions in cryptography. IEEE Trans Inf Theory IT-22(6):644–654
6. Schneier B (1998) Cryptographic design vulnerabilities. IEEE Comp 31(9).29–33
7. Rivest RL, Shamir A, Adleman L (1978) A method for obtaining digital signatures and public-key cryptosystems. Commun ACM 21(2):120–126

8. Wyner AD (1975) The wire-tap channel. Bell Syst Tech J 54(8):1355–1387
9. Leung-Yan-Cheong SK, Hellman ME (1978) The gaussian wire-tap channel. IEEE Trans Inf
    Theory IT-24(4):451–456
10. Csiszàr I, Körner J (1978) Broadcast channels with confidential messages. IEEE Trans Inf
    Theory 24(3):339–348
11. Ren K, Su H, Wang Q (2011) Secret key generation exploiting channel characteristics in
    wireless communications. IEEE Wirel Commun 18(4):6–12
12. Hero AO (2003) Secure space-time communication. IEEE Trans Inf Theory 49(12):3235–3249
13. Liang Y, Poor HV, Shamai (Shitz) S (2008) Secure communication over fading channels. IEEE
    Trans Inf Theory 54(6):2470–2492
14. Gopala PK, Lai L, El Gamal H (2008) On the secrecy capacity of fading channels. IEEE Trans
    Inf Theory 54(10):4687–4698
15. Khisti A, Wornell G (2010) Secure transmission with multiple antennas I: the MISOME wiretap
    channel. IEEE Trans Inf Theory 56(7):3088–3104
16. Khisti A, Wornell G (2010) Secure transmission with multiple antennas II: the MIMOME
    wiretap channel. IEEE Trans Inf Theory 56(11):5515–5532
17. Oggier F, Hassibi B (2011) The secrecy capacity of the MIMO wiretap channel. IEEE Trans
    Inf Theory 57(8):4961–4972
18. Bustin R, Liu R, Poor HV, Shamai (Shitz) S (2009) An MMSE approach to the secrecy capacity
    of the MIMO Gaussian wiretap channel. EURASIP J Wirel Commun Netw 2009
19. Maurer U (1993) Secret key agreement by public discussion from common information. IEEE
    Trans Inf Theory 39:733–742
20. Maurer U, Wolf S (2003) Secret-key agreement over unauthenticated public channels. IEEE
    Trans Inf Theory 49:822–838
21. Hassan AA, Stark WE, Hershey JE, Chennakeshu S (1996) Cryptographic key agreement for
    mobile radio. In: Signal digital processing, vol 6. Academic, San Diego, pp 207–212
22. Azimi-Sadjadi B, Mercado A, Kiayias A, Yener B (2007) Robust key generation from sig-
    nal envelopes in wireless networks. In: Proceedings of ACM computer and communications
    security, pp 401–410
23. Jana S, Premnath SN, Clark M, Kasera S, Patwari N, Krishnamurthy SV (2009) On the effective-
    ness of secret key extraction from wireless signal strength in real environments. In: Proceedings
    of ACM international conference on mobile computing and networking
24. Wilson R, Tse D, Scholtz RA (2007) Channel identification: secret sharing using reciprocity
    in ultrawideband channels. IEEE Trans Inf Forensics Secur 2:364–375
25. Wang Q, Su H, Ren K, Kim K (2011) Fast and scalable secret key generation exploiting channel
    phase randomness in wireless networks. In:Proceedings of IEEE International Conference on
    Computer Communications (INFOCOM), 2011
26. Dillard RA (1979) Detectability of spread-spectrum signals. IEEE Trans Aerosp Electron Syst
    AES-15(4):526–537
27. Gutman LL, Prescott GE (1989) System quality factors for LPI communication. IEEE Aerosp
    Electron Syst Mag 4(12):25–28
28. Flikkema P (1997) Spread-spectrum techniques for wireless communication. IEEE Signal
    Process Mag 14(3):26–36
29. Pickholtz RL, Schilling DL, Milstein LB (1982) Theory of spread-spectrum communications—
    a tutorial. IEEE Trans Commun 30(5):855–884
30. Kohno R, Meidan R, Milstein LB (1995) Spread spectrum access methods for wireless com-
    munications. IEEE Commun Mag 33(1):58–67
31. Spellman M (1983) A comparison between frequency hopping and direct spread PN as antijam
    techniques. IEEE Commun Mag 21(2):37–42
32. Burgos-Garcia M, Sanmartin-Jara J, Perez-Martinez F, Retamosa JA (2000) Radar sensor using
    low probability of interception SS-FH signals. IEEE Aerosp Electron Syst Mag 15(4):23–28
33. Liang Y, Kramer G, Poor HV, Shamai (Shitz) S (2009) Compound wiretap channels. EURASIP
    J Wirel Commun Netw 2009:5:1–5:12

34. Goel S, Negi R (2008) Guaranteeing secrecy using artificial noise. IEEE Trans Wirel Commun 7(6):2180–2189
35. Dong L, Han Z, Petropulu A, Poor H (2010) Improving wireless physical layer security via cooperating relays. IEEE Trans Signal Process 58(3):1875–1888
36. Huang J, Swindlehurst A (2012) Robust secure transmission in MISO channels based on worst-case optimization. IEEE Trans Signal Process 60(4):1696–1707
37. He X, Yener A (2010) Cooperation with an untrusted relay: a secrecy perspective. IEEE Trans Inf Theory 56(8):3807–3827
38. Jeong C, Kim I-M, Kim DI (2012) Joint secure beamforming design at the source and the relay for an amplify and forward MIMO untrusted relay system. IEEE Trans Signal Process 60(1):310–325
39. Chang T-H, Chiang W-C, Hong Y-WP, Chi C-Y (2010) Training sequence design for discriminatory channel estimation in wireless MIMO systems. IEEE Trans Signal Process 58(12):6223–6237
40. Huang C-W, Chang T-H, Zhou X, Hong Y-WP (2013) Two-way training for discriminatory channel estimation in wireless MIMO systems. IEEE Trans Signal Process 61(10):2724–2738

Chapter 2
Background on Information-Theoretic Physical Layer Secrecy

Abstract This chapter provides a brief overview of information-theoretic physical layer secrecy, including the fundamental limits and the key aspects that may affect the secrecy performance. The measure of secrecy as well as the notions of secrecy capacity and secrecy outage probability are introduced. Explicit expressions for the secrecy capacity and the secrecy outage probability are given for certain scenarios. These results motivate the development of the signal processing techniques to be introduced in later chapters.

Keywords Wiretap channel · Secrecy rate · Secrecy capacity · Secrecy outage · Channel code

Physical layer secrecy was first studied in the context of wiretap channels by Wyner in [1] and later on by Csiszár and Körner in [2]. In particular, a basic wiretap channel (or physical layer secrecy channel) consists of a source, a destination, and an eavesdropper that attempts to intercept the communication between the source and the destination. In the literature on information-theoretic physical layer secrecy, one is often interested in the maximum rate achievable between the source and the destination under the constraint on the information obtained by the eavesdropper (i.e., the secrecy constraint). This leads to the notions of secrecy rate and, secrecy capacity. The equivocation rate was also defined in the literature as an information-theoretic measure of the uncertainty of the confidential message at the eavesdropper. These studies proved the existence of secrecy coding or transmission schemes that are provably secure and provided results on the fundamental limits of the achievable rate under secrecy constraints. A brief summary of the information-theoretic results on physical layer secrecy is given in this chapter. These studies led to the development of the secrecy-enhancing signal processing techniques to be introduced in later chapters.

Y.-W. P. Hong et al., *Signal Processing Approaches to Secure Physical Layer*   11
Communications in Multi-Antenna Wireless Systems, SpringerBriefs in Signal Processing,
DOI: 10.1007/978-981-4560-14-6_2, © The Author(s) 2014

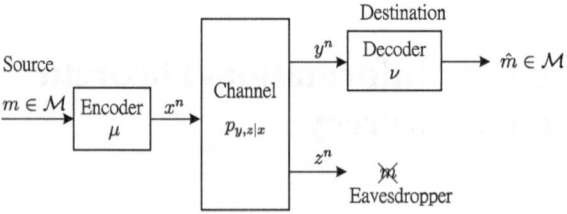

Fig. 2.1 The basic discrete memoryless wiretap channel

2.1 The Basic Wiretap Channel

Let us consider a basic discrete memoryless wiretap channel that consists of a source, a destination, and an eavesdropper, as shown in Fig. 2.1. The channel input at the source is denoted by the random variable x and the channel outputs at the destination and the eavesdropper are denoted by y and z, respectively. The random variables x, y, and z are drawn from the alphabets $\mathcal{X}$, $\mathcal{Y}$, and $\mathcal{Z}$, respectively. The channel input and output relations can be modeled generally by the conditional probability $p_{y,z|x}$, which describes the probability that the channel outputs at the destination and the eavesdropper are y and z given that the channel input is x.

Suppose that a confidential message m chosen from the message set $\mathcal{M} \triangleq \{1, \ldots, 2^{nR_s}\}$ is to be transmitted to the destination over n channel uses. Let $x^n \triangleq [x_1, \ldots, x_n]$ be the channel input over the n channel uses and let $y^n \triangleq [y_1, \ldots, y_n]$ and $z^n \triangleq [z_1, \ldots, z_n]$ be the corresponding channel outputs at the destination and the eavesdropper, respectively. A $(2^{nR_s}, n)$ wiretap code consists of a (stochastic) encoder $\mu : \mathcal{M} \to \mathcal{X}^n$ that maps a message $m \in \mathcal{M}$ into a length-n codeword $x^n \in \mathcal{X}^n$ and a decoder $\nu : \mathcal{Y}^n \to \mathcal{M}$ that maps the received sequence $y^n \in \mathcal{Y}^n$ to an estimated message $\hat{m} \in \mathcal{M}$. Note that the stochastic encoder maps each message $m \in \mathcal{M}$ randomly to a codeword $x^n \in \mathcal{X}^n$ according to a set of conditional probabilities, i.e., $\{p_{x^n|m}, \forall m \in \mathcal{M}, x^n \in \mathcal{X}^n\}$. The reception performance at the destination is measured by the average error probability defined as

$$P_e^{(n)} \triangleq \frac{1}{2^{nR_s}} \sum_{m=1}^{2^{nR_s}} \sum_{x^n \in \mathcal{X}^n} \Pr\left(\nu(y^n) \neq m | x^n\right) p_{x^n|m}. \qquad (2.1)$$

The level of secrecy at the eavesdropper is measured by the so-called equivocation rate, which is defined as the conditional entropy of m given z^n averaged over the number of channel uses, i.e.,

$$\frac{1}{n} H(m|z^n). \qquad (2.2)$$

The equivocation rate represents the eavesdropper's uncertainty about the message m given the channel outputs z^n. The larger the equivocation rate, the higher the level of secrecy that is achieved.

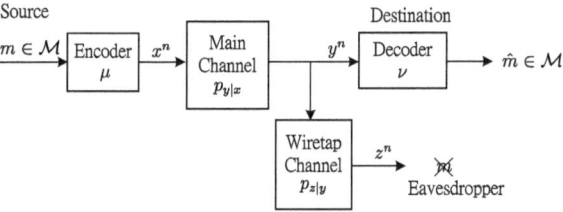

Fig. 2.2 The degraded wiretap channel

Definition 2.1 *A rate-equivocation pair* (R_s, R_e) *is said to be achievable if, for any* $\epsilon \geq 0$, *there exists an integer* $n'(\epsilon)$ *and a sequence of* $(2^{nR_s}, n)$-*codes such that, for all* $n \geq n'(\epsilon)$, *the average error probability is smaller than* ϵ, *i.e.,*

$$P_e^{(n)} \leq \epsilon,$$

and the equivocation rate is no less than $R_e - \epsilon$, *i.e.,*

$$\frac{1}{n} H(m|z^n) \geq R_e - \epsilon.$$

The rate-equivocation region $\mathcal{R}$ is defined as the set of all achievable rate-equivocation pairs (R_s, R_e). The rate-equivocation region $\mathcal{R}$ was first characterized for the degraded wiretap channel in [1] and was later extended to the more general non-degraded wiretap channels in [2]. In particular, the degraded wiretap channel refers to the case where the channel input and outputs satisfy the Markov relation $x \rightarrow y \rightarrow z$, as illustrated in Fig. 2.2. The result of the general non-degraded wiretap channel is given below.

Theorem 2.1 ([2]) *The rate-equivocation region of the wiretap channel is given by*

$$\mathcal{R} = \bigcup_{p_u, p_{v|u}, p_{x|v}} \left\{ (R_s, R_e) : \begin{array}{c} 0 \leq R_e \leq I(v; y|u) - I(v; z|u) \\ and \ R_e \leq R_s \leq I(v; y) \end{array} \right\} \tag{2.3}$$

where $I(x; y)$ *represents the mutual information between x and y, and u and v are auxiliary random variables that satisfy the Markov relation* $u \rightarrow v \rightarrow x \rightarrow (y, z)$.

In particular, many works in the literature are interested in the perfect secrecy scenario where the uncertainty at the eavesdropper should be equal to the randomness of the message, i.e., $R_e = R_s$. The maximum achievable secrecy rate in this case is referred to as the (perfect) secrecy capacity and can be derived as a special case of Theorem 2.1 by setting $R_e = R_s$.

Corollary 2.1 ([2]) *The (perfect) secrecy capacity of the wiretap channel is given by*

$$C_s = \max_{p_v, p_{x|v}} [I(v; y) - I(v; z)] \tag{2.4}$$

where v is an auxiliary random variable that satisfies the Markov relation $v \to x \to (y, z)$.

Specifically, by setting $R_e = R_s$ in Theorem 2.1, the secrecy capacity (i.e., the maximum achievable secrecy rate) can first be written as

$$C_s = \max_{p_u, p_{v|u}, p_{x|v}} [I(v; y|u) - I(v; z|u)], \tag{2.5}$$

where the maximization is taken over the distributions of u, v, and x that satisfy the Markov relation $u \to v \to x \to (y, z)$. Notice that the term inside the maximization satisfies

$$I(v; y|u) - I(v; z|u) = \sum_{\tilde{u} \in \mathcal{U}} \left[I(v; y|u = \tilde{u}) - I(v; z|u = \tilde{u}) \right] p_u(\tilde{u}) \tag{2.6}$$

$$\leq \max_{\tilde{u} \in \mathcal{U}} \left[I(v; y|u = \tilde{u}) - I(v; z|u = \tilde{u}) \right] \tag{2.7}$$

$$\leq \left[I(v^*; y) - I(v^*; z) \right], \tag{2.8}$$

where $\mathcal{U}$ is the alphabet corresponding to the random variable u, and v^* is the auxiliary random variable with distribution that maximizes (2.4). The upper bound in (2.8) is achieved by choosing $p_{v|u=\tilde{u}} = p_{v^*}$, for all $\tilde{u}$, and the secrecy capacity expression in (2.4) thus follows.

From (2.4), we can see that a positive secrecy rate can always be achieved, except for the case where $I(v; z) \geq I(v; y)$, for every auxiliary random variable v such that $v \to x \to (y, z)$, i.e., the case where the eavesdropper channel is *less noisy* than the main channel. It is worthwhile to remark that, to achieve the secrecy capacity, the original channel input is prefixed with the channel $p_{x|v}$, in which case, the auxiliary random variable v can be viewed as the effective channel input. This technique is referred to as *channel prefixing* [2].

An illustration of the rate-equivocation region is given in Fig. 2.3. We can see that perfect secrecy (i.e., the case where $R_e = R_s$) can be achieved when R_s is

Fig. 2.3 Illustration of the rate-equivocation region

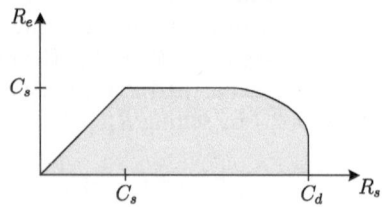

less than the secrecy capacity C_s. When $R_s > C_s$, the achievable equivocation rate R_e decreases monotonically with the increase of R_s. That is, the uncertainty at the eavesdropper reduces even though the uncertainty of the message increases.

Interestingly, the secrecy capacity expression can be further simplified for three special cases: the degraded, the less noisy, and the more capable wiretap channels [2]. Specifically, as mentioned previously, a wiretap channel is said to be *degraded* if the channel input and output variables satisfy the Markov relation $x \rightarrow y \rightarrow z$, i.e., when the output at the eavesdropper is a degraded version of that at the destination. Moreover, a wiretap channel is said to be *less noisy* if $I(v; y) \geq I(v; z)$, for every auxiliary random variable v such that $v \rightarrow x \rightarrow (y, z)$, and is said to be *more capable* if $I(x; y) \geq I(x; z)$, for every channel input x. One can show that the degraded condition is stronger than the less noisy condition, which is in order stronger than the more capable condition. That is, a wiretap channel is more capable if it is less noisy and is less noisy if it is degraded. When the wiretap channel is *more capable*, the secrecy capacity expression can be further simplified as follows:

$$C_s = \max_{p_v, p_{x|v}} [I(v; y) - I(v; z)] \tag{2.9}$$

$$= \max_{p_v, p_{x|v}} \{I(x; y) - I(x; z) - [I(x; y|v) - I(x; z|v)]\} \tag{2.10}$$

$$= \max_{p_x} [I(x; y) - I(x; z)]. \tag{2.11}$$

This follows from the fact that $I(x; y|v) - I(x; z|v) \geq 0$ due to the more capable condition and can be made equal to 0 by setting $v = x$. Since both the less noisy and the degraded conditions imply the more capable condition, the secrecy capacity formula given in (2.11) holds for these two cases as well.

It is interesting to remark that the secrecy capacity of the non-degraded wiretap channel, as given in (2.4), was derived by Csiszár and Körner in [2] as a special case of the more general broadcast channel with confidential messages. In the latter case, two messages are to be transmitted, namely, a common message t that is to be decoded by both the destination and the eavesdropper, and a private message m that is to be decoded only by the destination. A rate-equivocation triplet (R_s, R_t, R_e) is said to be achievable if both messages can be decoded at the destination and the common message can be decoded at the eavesdropper with arbitrarily low error probability, and the equivocation rate of the private message at the eavesdropper is greater than R_e. The rate-equivocation region in this case is given as follows.

Theorem 2.2 ([2]) *The rate-equivocation region of the broadcast channel with confidential message is given by*

$$\mathcal{R} = \bigcup_{p_u, p_{v|u}, p_{x|v}} \left\{ \begin{array}{l} (R_s, R_t, R_e): \\ 0 \leq R_e \leq R_s, \quad R_e \leq I(v; y|u) - I(v; z|u), \\ R_s + R_t \leq I(v; y|u) + \min[I(u; y), I(u, z)], \\ 0 \leq R_t \leq \min[I(u; y), I(u; z))] \end{array} \right\} \tag{2.12}$$

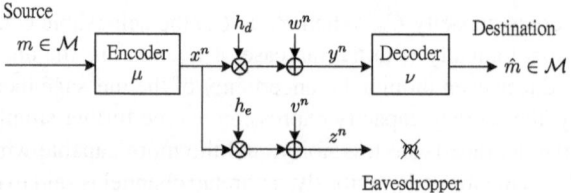

Fig. 2.4 The Gaussian wiretap channel

where u and v are auxiliary random variables that satisfy the Markov relation $u \rightarrow v \rightarrow x \rightarrow (y, z)$.

When no common message is to be transmitted (i.e. when $R_t = 0$), the rate-equivocation region reduces to the form in Theorem 2.1.

In the literature, studies on the rate-equivocation region and the secrecy capacity have also been extended to other scenarios, such as the interference channel with confidential messages [3, 4], the broadcast channel with different confidential messages to the destinations [3, 5], and the multiple access wiretap channel [6, 7]. However, the rate-equivocation region and/or the secrecy capacity often cannot be characterized explicitly, in which case, inner and outer bounds are derived. Readers are referred to [8] for a more detailed survey of these topics. It is also worthwhile to remark that, in proving the achievability of the above results, the random coding arguments are used to show the existence of good channel codes that can simultaneously achieve a low error probability at the destination while maintaining perfect secrecy against an eavesdropper. However, random codes are not applicable in practice and more practical codes are needed. Readers are referred to [9–11] for studies on the design of practical wiretap codes.

2.2 Gaussian and MIMO Gaussian Wiretap Channels

In this section, secrecy capacity expressions are given for the Gaussian and the multiple-input multiple-output (MIMO) Gaussian wiretap channels. Specifically, let us first consider a Gaussian wiretap channel, as shown in Fig. 2.4, where the channel output at the destination and the eavesdropper are corrupted by additive white Gaussian noise (AWGN). Let x represent the channel input and let y and z represent the corresponding channel outputs at the destination and the eavesdropper, respectively. The channel input and output relations can be described as follows:

$$y = h_d x + w \qquad (2.13a)$$
$$z = h_e x + v \qquad (2.13b)$$

where h_d and h_e are the channel coefficients and $w \sim \mathcal{CN}(0, \sigma_w^2)$ and $v \sim \mathcal{CN}(0, \sigma_v^2)$ are the AWGN at the destination and the eavesdropper, respectively. The codeword transmitted over n channel uses, i.e., $x^n = [x_1, \ldots, x_n]$, must satisfy the average power constraint

$$\frac{1}{n} \sum_{i=1}^{n} E[|x_i|^2] \leq \bar{P} \tag{2.14}$$

where $\bar{P}$ is the maximum average power. The AWGN is assumed to be independent and identically distributed (i.i.d.) over time. By viewing h_d and h_e as the fading coefficients, the signal model given in (2.13) can also be viewed as the signal model for the single-input single-output (SISO) wireless system.

The secrecy capacity of the Gaussian wiretap channel was derived in [12] and is given below.

Theorem 2.3 ([12]) *The secrecy capacity of the Gaussian wiretap channel is given by*

$$C_s = \left[\log \left(1 + \frac{|h_d|^2 \bar{P}}{\sigma_w^2} \right) - \log \left(1 + \frac{|h_e|^2 \bar{P}}{\sigma_v^2} \right) \right]^+ \tag{2.15}$$

where $[\cdot]^+ = \max(0, \cdot)$ and $\bar{P}$ is the average power constraint.

The Gaussian wiretap channel can be viewed as a (stochastically) degraded wiretap channel when $|h_d|^2/\sigma_w^2 > |h_e|^2/\sigma_v^2$. Therefore, similar to that in (2.11), the secrecy capacity given in (2.15) can be achieved by setting $u = x$ and by letting x be $\mathcal{CN}(0, \bar{P})$. When $|h_d|^2/\sigma_w^2 \leq |h_e|^2/\sigma_v^2$, the main channel becomes a degraded version of the eavesdropper channel and hence no positive secrecy rate can be obtained.

The Gaussian wiretap channel described above can also be extended to systems with multiple antennas at each terminal. Specifically, let us consider a multiple-input multiple-output (MIMO) Gaussian wiretap channel, as illustrated in Fig. 2.5, where the source, the destination, and the eavesdropper are equipped with n_s, n_d and n_e antennas respectively. Let $\mathbf{x} = [x_1, \ldots, x_{n_s}]^T$ represent the channel inputs across the n_s transmit antennas and let $\mathbf{y} = [y_1, \ldots, y_{n_d}]^T$ and $\mathbf{z} = [z_1, \ldots, z_{n_e}]^T$ be the channel outputs at the destination and the eavesdropper, respectively. The channel input and output relations can be described as follows:

Fig. 2.5 The MIMO Gaussian wiretap channel

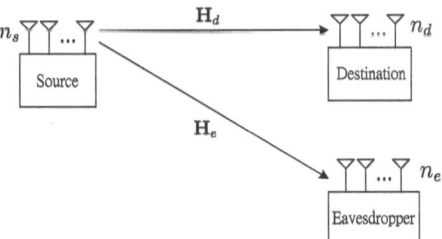

$$\mathbf{y} = \mathbf{H}_d \mathbf{x} + \mathbf{w}, \tag{2.16a}$$

$$\mathbf{z} = \mathbf{H}_e \mathbf{x} + \mathbf{v}, \tag{2.16b}$$

where $\mathbf{H}_d \in \mathbb{C}^{n_d \times n_s}$ and $\mathbf{H}_e \in \mathbb{C}^{n_e \times n_s}$ are the channel matrices corresponding to the main and the eavesdropper channels, respectively, and $\mathbf{w} \in \mathbb{C}^{n_d \times 1}$ and $\mathbf{v} \in \mathbb{C}^{n_e \times 1}$ are the AWGN vectors at the destination and the eavesdropper, respectively. The AWGN vectors are assumed to have i.i.d. entries with zero-mean and unit-variance, i.e., $\mathbf{w} \sim \mathcal{CN}(0, \mathbf{I}_{n_d})$ and $\mathbf{v} \sim \mathcal{CN}(0, \mathbf{I}_{n_e})$. The channel matrices $\mathbf{H}_d$ and $\mathbf{H}_e$ are assumed to be known at all terminals and remain constant over the transmission of a codeword. Let $\mathbf{x}^n = [\mathbf{x}_1, \ldots, \mathbf{x}_n]$ be the transmitted codeword, where $\mathbf{x}_i$ is the $n_s \times 1$ channel input vector during the ith channel use. The codeword transmitted over n channel uses must satisfy the average power constraint given by

$$\mathrm{E}\left[\frac{1}{n}\sum_{i=1}^{n}||\mathbf{x}_i||^2\right] \le \bar{P}, \tag{2.17}$$

where $\bar{P}$ is the maximum average power. The secrecy capacity of the MIMO Gaussian wiretap channel was derived in [13–15] and is given below.

Theorem 2.4 ([13–15]) *The secrecy capacity of the MIMO Gaussian wiretap channel is given by*

$$C_s = \max_{\mathbf{K_x} \succeq \mathbf{0}, \mathrm{tr}(\mathbf{K_x}) \le \bar{P}} \log \frac{\det(\mathbf{I}_{n_d} + \mathbf{H}_d \mathbf{K_x} \mathbf{H}_d^H)}{\det(\mathbf{I}_{n_e} + \mathbf{H}_e \mathbf{K_x} \mathbf{H}_e^H)}, \tag{2.18}$$

where $\mathbf{K_x} \triangleq \mathrm{E}[\mathbf{x}\mathbf{x}^H]$ is the input covariance matrix.

The secrecy capacity expression in (2.18) is achieved by taking $u = \mathbf{x}$ and by letting $\mathbf{x} \sim \mathcal{CN}(\mathbf{0}, \mathbf{K_x})$. The converse however is based on upper bounds that are derived by considering fictitious degraded wiretap channels, where the channel output at the eavesdropper is also revealed to the destination. However, the secrecy capacity expression in (2.18) is non-convex in $\mathbf{K_x}$, which poses challenges in determining explicitly the optimal input covariance matrix. Yet, as shown in [16] and described in the following chapter, explicit solutions can be obtained when power covariance constraints are considered instead of the average sum power constraint.

2.3 Compound Wiretap Channel

The basic and Gaussian wiretap channels introduced in previous sections can also be extended to the case with multiple destinations and multiple eavesdroppers. This can be viewed as a special case of the so-called *compound wiretap channel* [17]. In

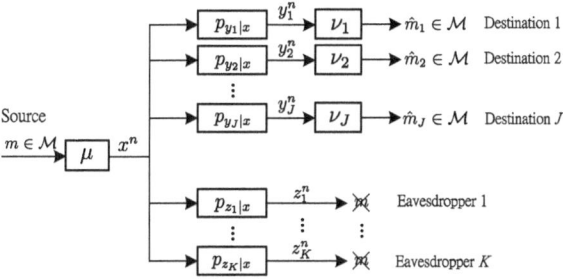

Fig. 2.6 The compound wiretap channel

particular, the compound wiretap channel models are more generally the case where the main and the eavesdropper channels can each take a number of different states and that confidentiality must be achieved regardless of which pair of channel states occurs.

Let us consider a compound wiretap channel that consists of a source, J destinations, and K eavesdroppers, as illustrated in Fig. 2.6. Let $x \in \mathcal{X}$ be the channel input and let $y_j \in \mathcal{Y}_j$ and $z_k \in \mathcal{Z}_k$ be the channel output at the jth destination and the kth eavesdropper, where $\mathcal{X}$, $\mathcal{Y}_j$, for $j = 1, \ldots, J$, and $\mathcal{Z}_k$, for $k = 1, \ldots, K$, are the channel input and output alphabets. The channel input and output relations are described by the conditional probabilities $p_{y_j|x}$, for $j = 1, \ldots, J$, and $p_{z_k|x}$, for $k = 1, \ldots, K$. A $(2^{nR_s}, n)$ code consists of a (stochastic) encoder $\mu : \mathcal{M} \to \mathcal{X}^n$, where $\mathcal{M} \triangleq \{1, \ldots, 2^{nR_s}\}$ is the message, that maps the message into a length-n codeword and J decoders $\nu_j : \mathcal{Y}_j^n \to \mathcal{M}$, for $j = 1, \ldots, J$, that maps the received signal at each destination to a message estimate in $\mathcal{M}$. The average error probability for destination j is given by

$$P_{e,j}^{(n)} \triangleq \frac{1}{2^{nR_s}} \sum_{m=1}^{2^{nR_s}} \sum_{x^n \in \mathcal{X}^n} \Pr\left(\nu_j(y_j^n) \neq m | x^n\right) p_{x^n|m} \tag{2.19}$$

and the equivocation rate at eavesdropper k is given by

$$\frac{1}{n} H(m|z_k^n). \tag{2.20}$$

A secrecy rate R_s is achievable if, for any $\epsilon \geq 0$, there exists $n'(\epsilon)$ and a sequence of $(2^{nR_s}, n)$ codes such that, for all $n \geq n'(\epsilon)$, the average error probabilities at the J destinations are all less than ϵ, i.e.,

$$P_{e,j}^{(n)} \leq \epsilon,$$

for $j = 1, \ldots, J$, and the equivocation rate at the K eavesdroppers are all ϵ close to R_s, i.e.,

$$\frac{1}{n} H(m|z_k^n) \geq R_s - \epsilon,$$

for $k = 1, \ldots, K$. The secrecy capacity is the maximum achievable secrecy rate. The secrecy capacity of the compound wiretap channel is not known in general, but lower and upper bounds were derived in [17], as summarized below.

Theorem 2.5 ([17]) *The secrecy capacity of the compound wiretap channel is lower bounded by the achievable secrecy rate*

$$R_{s,\text{lower}} = \max_{p_u \cdot p_{x|u}} \left\{ \min_{j,k} \left[I(u; y_j) - I(u; z_k) \right] \right\}, \tag{2.21}$$

where the maximization is taken over the distributions of u and x that satisfy $u \to x \to (y_j, z_k)$, for $j = 1, \ldots, J$ and $k = 1, \ldots, K$, and is upper bounded by

$$R_{s,\text{upper}} = \min_{j,k} \left\{ \max_{p_u \cdot p_{x|u}} \left[I(u; y_j) - I(u; z_k) \right] \right\}. \tag{2.22}$$

The upper bound is not achievable in general since the distributions of u and x may not be simultaneously optimized for all destination-eavesdropper pairs. The lower bound can be viewed as the secrecy rate achievable in a wiretap channel consisting of the worst destination and the best eavesdropper. The secrecy capacity is exactly equal to the lower bound when the compound wiretap channel is degraded, i.e., when the channel input and outputs satisfy the Markov relations $x \to y_j \to z_k$, for all $j = 1, \ldots, J$, and $k = 1, \ldots, K$. In this case, the secrecy capacity is given by

$$C_s = \max_{p_x} \left\{ \min_{j,k} \left[I(x; y_j) - I(x; z_k) \right] \right\}, \tag{2.23}$$

which is achieved by setting $u = x$.

The secrecy capacity can also be derived for the Gaussian compound wiretap channel with $J = 1$ as shown in [18]. In this case, the received signal at the destination and the eavesdroppers are given by

$$y = h_d x + w, \tag{2.24}$$
$$z_k = h_{e,k} x + v_k, \quad \text{for } k = 1, \ldots, K, \tag{2.25}$$

where $w \sim \mathcal{CN}(0, \sigma_w^2)$ and $v_k \sim \mathcal{CN}(0, \sigma_{v,k}^2)$ are the AWGN at the destination and the kth eavesdropper, respectively. The secrecy capacity is given by [18]

$$C_s = \min_k \left[\log \left(1 + \frac{P|h_d|^2}{\sigma_w^2} \right) - \log \left(1 + \frac{P|h_{e,k}|^2}{\sigma_{v,k}^2} \right) \right]^+. \tag{2.26}$$

The compound wiretap channel can also be extended to the MIMO case. Specifically, let us consider the MIMO Gaussian compound wiretap channel where the channel input and output relations are given by

$$\mathbf{y}_j = \mathbf{H}_{d,j}\mathbf{x} + \mathbf{w}_j, \tag{2.27}$$

$$\mathbf{z}_k = \mathbf{H}_{e,k}\mathbf{x} + \mathbf{v}_k, \tag{2.28}$$

for $j = 1, \ldots, J$ and $k = 1, \ldots, K$, where $\mathbf{w}_j \sim \mathcal{CN}(\mathbf{0}, \mathbf{I})$ and $\mathbf{v}_k \sim \mathcal{CN}(\mathbf{0}, \mathbf{I})$. The secrecy capacity is not known in general for this scenario, but an achievable secrecy rate can be given by [19]

$$R_s = \max_{\mathbf{K_x} \succeq 0, \mathrm{tr}(\mathbf{K_x}) \leq \bar{P}} \min_{j,k} \log \frac{\det(\mathbf{I}_{n_d} + \mathbf{H}_{d,j}\mathbf{K_x}\mathbf{H}_{d,j}^H)}{\det(\mathbf{I}_{n_e} + \mathbf{H}_{e,k}\mathbf{K_x}\mathbf{H}_{e,k}^H)}, \tag{2.29}$$

Notice that this is analogous to the achievable rate in Theorem 2.5 which can be viewed as the worst case rate, i.e., the secrecy rate of the wiretap channel consisting of the worst destination and the best eavesdropper.

2.4 Ergodic Secrecy Capacity

The secrecy capacity expressions given in the previous sections were derived for cases where channels remain static over the transmission of a codeword. It was shown, in those cases, that a non-zero secrecy rate can be achieved only when the main channel is more favorable than the eavesdropper channel. However, in the wireless fading environment, channel coefficients may vary drastically over different dimensions in time, frequency, and space. In this case, the source can exploit opportunistically the dimensions for which the main channel is better than the eavesdropper channel and, in this way, achieve a positive secrecy rate even when the eavesdropper channel is more favorable *on the average*. This is achieved by coding over a long period of time and over multiple channel states. The maximum achievable secrecy rate in this case is referred to as the *ergodic secrecy capacity* .

Let us consider a SISO wiretap channel where the source, the destination, and the eavesdropper are each equipped with only a single antenna. The signal received at the destination and the eavesdropper are given by

$$y = h_d x + w, \tag{2.30a}$$

$$z = h_e x + v, \tag{2.30b}$$

where h_d and h_e are the main and the eavesdropper channel coefficients and $w \sim \mathcal{CN}(0, \sigma_w^2)$ and $v \sim \mathcal{CN}(0, o_v^2)$ are the AWGN at the destination and the eavesdropper, respectively. The channel coefficients h_d and h_e are assumed to remain constant over each coherence interval and vary independently over different intervals. Each coherence interval is assumed to be large enough such that the destination can successfully decode under random coding arguments. When the destination's and the eavesdropper's channel state information (CSI) is available at the source, a non-zero ergodic secrecy capacity can be achieved, as shown in [19, 20], by allocating rate and power only over the channel states for which the main channel is more favorable than the eavesdropper channel. The ergodic secrecy rate is summarized in the following theorem.

Theorem 2.6 ([20]) *Given knowledge of the destination's and the eavesdropper's CSI at the source, the ergodic secrecy capacity of the slow fading wiretap channel is given by*

$$
C_s = \max_{P(\cdot,\cdot) \in \mathcal{P}} \mathrm{E}\left[\left[\log\left(1 + \frac{P(h_d, h_e)|h_d|^2}{\sigma_w^2}\right) - \log\left(1 + \frac{P(h_d, h_e)|h_e|^2}{\sigma_v^2}\right) \right]^+ \right]
$$

(2.31)

where $P(h_d, h_e)$ is a power allocation function that takes both the main and the eavesdropper channel coefficients as the input and $\mathcal{P} \triangleq \{P(\cdot, \cdot) : \mathrm{E}[P(h_d, h_e)] \le \bar{P}\}$ is the set of power allocation functions that satisfy the power constraint $\bar{P}$. The optimal power allocation function is given by

$$
P^*(h_d, h_e) =
$$

$$
\begin{cases}
\left(\dfrac{1}{\lambda \ln 2} - \dfrac{\sigma_w^2}{|h_d|^2} \right)^+, & \text{if } |h_d|^2 > 0 \text{ and } |h_e|^2 = 0 \\[2ex]
\dfrac{1}{2}\left[\sqrt{\left(\dfrac{\sigma_v^2}{|h_e|^2} - \dfrac{\sigma_w^2}{|h_d|^2}\right)^2 + \dfrac{4}{\lambda \ln 2}\left(\dfrac{\sigma_v^2}{|h_e|^2} - \dfrac{\sigma_w^2}{|h_d|^2}\right)} - \left(\dfrac{\sigma_w^2}{|h_d|^2} + \dfrac{\sigma_v^2}{|h_e|^2}\right) \right]^+, \\[1ex]
& \text{if } |h_d|^2/\sigma_w^2 > |h_e|^2/\sigma_v^2 > 0 \\[1ex]
0, & \text{otherwise}
\end{cases}
$$

(2.32)

where λ is a constant chosen such that $\mathrm{E}[P^*(h_d, h_e)] = \bar{P}$.

The optimal power allocation function given in (2.32) shows that power should only be allocated to channels for which the main channel is sufficiently better than the eavesdropper channel (that is, the main channel SNR $|h_d|^2/\sigma_w^2$ is sufficiently larger than the eavesdropper channel SNR $|h_e|^2/\sigma_v^2$). In fact, the allocated power should increase as the difference in SNR inverse between the two channels (or the difference of its reciprocal i.e., $\sigma_v^2/|h_e|^2 - \sigma_w^2/|h_d|^2$) increases.

It is worthwhile to remark that the secrecy capacity expressions given in the previous sections and the ergodic secrecy capacity given above are all derived based on the assumption that both the main and the eavesdropper channel state information (CSI) are available at the source. However, this may not be the case in practice since the eavesdropper, as an adversary, would not want to reveal any information to the source that may weaken its ability to intercept the message. In the following, we consider the case where only the destination's CSI is available at the source. The ergodic secrecy capacity of this case was also derived in [20] and is given in the following theorem.

Theorem 2.7 ([20]) *Given knowledge of only the destination's CSI at the source, the ergodic secrecy capacity of the slow fading wiretap channel is given by*

$$C_s = \max_{P(\cdot) \in \mathcal{P}'} E\left[\left[\log\left(1 + \frac{P(h_d)|h_d|^2}{\sigma_w^2}\right) - \log\left(1 + \frac{P(h_d)|h_e|^2}{\sigma_v^2}\right)\right]^+\right] \quad (2.33)$$

where $P(\cdot)$ is the power allocation function that takes only the destination's CSI h_d as the input and $\mathcal{P}' \triangleq \{P(\cdot) : E[P(h_d)] \leq \bar{P}\}$.

Notice that, without the eavesdropper's CSI at the source, the source is only able to adapt its rate and power to the channel states of the source. Hence, the source may be transmitting even during time instants in which the eavesdropper is experiencing a better channel than the destination. However, a non-zero ergodic secrecy rate can still be achieved in this case by hiding the message across multiple channel states. In particular, this is achieved by employing a variable rate transmission scheme where, during a coherence interval of channel state h_d, the source transmits a codeword with rate $\log(1 + P(h_d)|h_d|^2/\sigma_w^2)$. By doing so, the mutual information between the source and the eavesdropper is upper bounded by $\log(1 + P(h_d)|h_d|^2/\sigma_w^2)$ when $|h_e|^2/\sigma_v^2 > |h_d|^2/\sigma_w^2$ and by $\log(1 + P(h_d)|h_e|^2/\sigma_v^2)$ when $|h_e|^2/\sigma_v^2 < |h_d|^2/\sigma_w^2$. By averaging over the channel states, the mutual information accumulated at the destination is $E[\log(1 + P(h_d)|h_d|^2/\sigma_w^2)]$ whereas that at the eavesdropper is $E[\log(1 + P(h_d) \min\{|h_d|^2/\sigma_w^2, |h_e|^2/\sigma_v^2\})]$. Hence, the achievable secrecy rate is thus given by the difference of the accumulated mutual information, i.e.,

$$E\left[\log\left(1 + P(h_d)\frac{|h_d|^2}{\sigma_w^2}\right) - \log\left(1 + P(h_d) \min\left\{\frac{|h_d|^2}{\sigma_w^2}, \frac{|h_e|^2}{\sigma_v^2}\right\}\right)\right], \quad (2.34)$$

which then leads to the expression in (2.33). A more rigorous proof can be found in [20].

Moreover, since the objective function in (2.33) is concave, the optimal power allocation function can be found by using the Lagrangian maximization approach [20]. This results in the following optimality condition

$$\frac{\partial C_s}{\partial P(h_d)} =$$

$$\frac{|h_d|^2 \Pr\left(\frac{|h_e|^2}{\sigma_v^2} \leq \frac{|h_d|^2}{\sigma_w^2} \Big| h_d\right)}{\sigma_w^2 + P(h_d)|h_d|^2} - E\left[\frac{|h_e|^2}{\sigma_v^2 + P(h_d)|h_e|^2} \mathbf{1}_{\{\frac{|h_e|^2}{\sigma_v^2} < \frac{|h_d|^2}{\sigma_w^2}\}} \Big| h_d\right] - \lambda = 0$$

where λ is chosen such that $E[P(h_d)] = \bar{P}$. Different from the case with full CSI at the source, the optimal power allocation function now requires averaging over the eavesdropper's channel coefficient and, thus, cannot be written in closed-form.

The above discussions on the ergodic secrecy capacity can also be extended to systems with multiple antennas at each terminal, e.g., in [21, 22].

2.5 Secrecy Outage

To achieve the ergodic secrecy capacity described in the previous section, it is necessary to encode over a long period of time and, thus, over all channel realizations. This typically incurs long delay and is suitable only for delay-tolerant applications. In this section, we focus on delay-limited applications, where the confidential message is encoded only over a single coherence interval or channel block. Each channel block is assumed to be sufficiently long so that small error probability can be achieved as long as the code-rate is below the capacity of the wiretap channel. In this case, we are concerned with the so-called *secrecy outage probability* [19, 23], which is the probability that the target secrecy rate cannot be achieved in a given channel block.

Specifically, let C_s be the instantaneous secrecy capacity in a given channel block (or coherence interval) and let R_0 be the target secrecy rate, i.e., the rate of the codeword. We say that a secrecy outage occurs when the secrecy capacity C_s falls below the target secrecy rate R_0. The secrecy outage probability is thus defined as follows.

Definition 2.2 *Secrecy outage probability is the probability that the secrecy capacity C_s falls below the target secrecy rate R_0, i.e.,*

$$P_{s,out}(R_0) = \Pr(C_s \leq R_0) \tag{2.35}$$

where R_0 is the target secrecy rate.

When both the destination's and the eavesdropper's CSI are available at the source, the source can determine when to transmit and when not to transmit with a fixed secrecy codebook. In this case, the secrecy outage probability represents the fraction of time that the source should remain silent (since, otherwise, confidentiality would be compromised with the fixed rate codebook). When the eavesdropper's CSI is unknown, confidentiality cannot be maintained in each channel block and hence, in this case, the secrecy outage probability measures only the fraction of time that secrecy is compromised.

Specifically, let us consider a SISO wiretap channel with the signal model given in (2.30). We again consider the block-fading scenario where the channel coefficients are assumed to remain constant over each block (or coherence interval) and vary independently from block to block. Different from the previous section, we assume here that each codeword is transmitted within a single channel block and assume that each block is sufficiently long such that a codeword can be transmitted with small error probability. Let us consider the Rayleigh fading example where h_d and h_e are assumed to be i.i.d. circularly symmetric complex Gaussian random variables with zero mean and variances $\sigma_{h_d}^2$ and $\sigma_{h_e}^2$, respectively. In this case, the SNR at the destination and the eavesdropper, i.e., $\Gamma_d \triangleq P|h_d|^2/\sigma_w^2$ and $\Gamma_e \triangleq P|h_e|^2/\sigma_v^2$, will be exponentially distributed with densities given by mean $\bar{\gamma}_d = P\sigma_{h_d}^2/\sigma_d^2$ and $\bar{\gamma}_e = P\sigma_{h_e}^2/\sigma_e^2$, respectively. Given the channel realizations h_d and h_e, the channel in (2.30) is equivalent to a Gaussian wiretap channel and, thus, the secrecy capacity can be written as

$$C_s(h_d, h_e) = \left[\log\left(1 + \frac{P|h_d|^2}{\sigma_w^2}\right) - \log\left(1 + \frac{P|h_e|^2}{\sigma_v^2}\right) \right]^+ \tag{2.36}$$

where $P \triangleq \mathrm{E}[|x|^2]$ is the average transmit power. Hence, the secrecy outage probability can be computed as

$$
\begin{aligned}
P_{s,out}(R_0) &= \Pr\left(C_s(h_d, h_e) \le R_0\right) \\
&= 1 - \Pr\left(\frac{1 + P|h_d|^2/\sigma_w^2}{1 + P|h_e|^2/\sigma_v^2} > 2^{R_0}\right) \\
&= 1 - \Pr\left(\Gamma_d > 2^{R_0}(1 + \Gamma_e) - 1\right) \\
&= 1 - \int_0^\infty \int_{2^{R_0}(1+\gamma_e)-1}^\infty \frac{1}{\bar{\gamma}_d} \exp\left(-\frac{\gamma_d}{\bar{\gamma}_d}\right) \cdot \frac{1}{\bar{\gamma}_e} \exp\left(-\frac{\gamma_e}{\bar{\gamma}_e}\right) d\gamma_d d\gamma_e \\
&= 1 - \frac{\bar{\gamma}_d}{\bar{\gamma}_d + 2^{R_0}\bar{\gamma}_e} \exp\left(-\frac{2^{R_0}-1}{\bar{\gamma}_d}\right).
\end{aligned}
\tag{2.37}
$$

One can observe from (2.37) that the secrecy outage probability decreases as the average SNR at the destination (i.e., $\bar{\gamma}_d$) increases but increases with the average SNR at the eavesdropper (i.e., $\bar{\gamma}_e$) as well as the target rate R_0.

With instantaneous knowledge of h_d and h_e are available at the source, the source can not only determine when to transmit and when not to transmit, but can more generally determine the optimal power allocation subject to a long-term power constraint, i.e., $\mathrm{E}[P(h_d, h_e)] \le \bar{P}$, where $P(h_d, h_e)$ is the power utilized under the channel state (h_d, h_e). In each channel block, the source can assign just enough power to achieve the target secrecy rate when it decides to transmit. In this case, the minimum power required to achieve target secrecy rate R_0 given h_d and h_e is [19]

$$P_{\min}(h_d, h_e) = \begin{cases} \dfrac{2^{R_0}-1}{|h_d|^2/\sigma_w^2 - 2^{R_0}|h_e|^2/\sigma_v^2}, & \text{if } R_0 < \log\dfrac{|h_d|^2/\sigma_w^2}{|h_e|^2/\sigma_v^2} \\ \infty, & \text{otherwise.} \end{cases} \tag{2.38}$$

However, to minimize outage probability under the long-term power constraint, the source should choose to transmit only in channels that require less power to achieve the target secrecy rate and remain silent in channels that require too much power. Hence, whether or not the source should transmit in a given channel should be a threshold decision on the amount of power required to reach the target secrecy rate. Hence, the optimal power allocation is given by

$$P(h_d, h_e) = P_{\min}(h_d, h_e)\mathbf{1}_{\{P_{\min}(h_d,h_e)\leq\lambda\}}, \tag{2.39}$$

where λ is chosen such that $E[P(h_d, h_e)] = \bar{P}$.

In addition to the secrecy outage probability, one may also be interested in the maximum target secrecy rate that can be chosen subject to a constraint on the secrecy outage probability. This is referred to as the secrecy outage capacity [24] as defined below.

Definition 2.3 *The ϵ-secrecy outage capacity is the maximum target secrecy rate that yields secrecy outage probability less than ϵ, i.e.,*

$$C_{s,out}(\epsilon) = \sup_{E[|x|^2]\leq\bar{P}} \sup\{R : P_{s,out}(R) \leq \epsilon\} \tag{2.40}$$

where the first supremum is taken over input distributions satisfying $E[|x|^2] \leq \bar{P}$ and the second supremum is taken over all target secrecy rates R.

A closed-form expression of secrecy outage capacity was derived in [25] for the single-input single-output single-antenna eavesdropper (SISOSE) channel described in (2.30) and is given by

$$C_{s,out}(\epsilon) = \log_2\left(\frac{1 + \epsilon\bar{\gamma}_d}{1 + (1-\epsilon)\bar{\gamma}_e}\right).$$

The concept of secrecy outage probability and secrecy outage capacity can also be extended to the case with multiple antennas at different terminals. Readers are referred to [24, 26] for further studies on this topic.

2.6 Summary and Discussions

In this chapter, we provided a brief summary of the basic results on information-theoretic physical layer secrecy. We first considered the basic wiretap channel that consists of a source, a destination, and an eavesdropper, who passively intercepts messages transmitted by the source. In the information theory literature, one is often interested in the so-called secrecy capacity, which is defined as the maximum rate achievable between the source and the destination subject to an arbitrarily low error probability at the destination and a lower bound on the equivocation rate at the eavesdropper. The equivocation rate is a measure of randomness or uncertainty at the eavesdropper. In particular, many works in the literature focus on the perfect secrecy scenario where the equivocation rate is to be lower bounded by the rate of the confidential message. That is, the entropy of the message given by the eavesdropper's channel output is the same as the message itself and, thus, no additional information can be obtained from the eavesdropper's channel output. The secrecy capacity was first given for the general discrete memoryless wiretap channel (in the form of an optimization problem) and was derived more explicitly for Gaussian and MIMO Gaussian wiretap channels. The latter will be more relevant to our discussions in later chapters. The secrecy capacity can be roughly viewed (subject to prefixing with an auxiliary random variable u) as the difference between the mutual information of the main channel and that of the eavesdropper channel. A non-zero secrecy rate can be achieved only when the destination experiences a better channel than the eavesdropper. Similar concepts were also introduced for the compound wiretap channel, which can be viewed as the case with multiple destinations and multiple eavesdroppers.

In addition to the non-ergodic secrecy capacity, the ergodic secrecy capacity and the secrecy outage probability were also introduced as performance measures often used to capture the impact of fading on the achievable physical layer secrecy. In particular, the ergodic secrecy capacity characterizes the maximum secrecy rate achievable when coding over a long period of time and across multiple channel states. It was shown that, by exploiting opportunistically the dimensions (e.g., time instants) for which the main channel is more favorable than the eavesdropper channel, a non-zero ergodic secrecy capacity can be achieved even when the channel condition at the eavesdropper is, on the average, more favorable than that at the destination. When the eavesdropper's CSI is unknown at the source, confidentiality can be maintained by hiding the message over multiple channel states, which prevents the eavesdropper from intercepting the message even when its channel may be temporarily better than that of the destination. For delay-limited applications, where the message must be transmitted over a single coherence interval, the secrecy outage probability provides a measure of the fraction of time that secrecy transmission can be successfully performed with a fixed rate codebook. When the eavesdropper's CSI is unknown at the source, the secrecy outage probability provides a measure of the probability that secrecy is violated.

The information-theoretic studies introduced in this chapter motivate the development of the signal processing approaches that are to be introduced in the following chapters. In particular, based on the secrecy capacity expressions, signal processing techniques can be developed to enlarge the difference in signal quality between the destination and the eavesdropper and, in this way, enhance the physical layer secrecy channel. A more detailed survey on information-theoretic secrecy can be found in [8].

References

1. Wyner AD (1975) The wire-tap channel. Bell Syst Tech J 54(8):1355–1387
2. Csiszàr I, Körner J (1978) Broadcast channels with confidential messages. IEEE Trans Inf Theory 24(3):339–348

3. Liu R, Maric I, Spasojević P, Yates RD (2008) Discrete memoryless interference and broadcast channels with confidential messages-secrecy rate regions. IEEE Trans Inf Theory 54(6):1–14
4. Liang Y, Somekh-Baruch A, Poor HV, Shamai S, Verdu S (2009) Capacity of cognitive interference channels with and without secrecy. IEEE Trans Inf Theory 55(2):604–619
5. Liu R, Poor HV (2009) Secrecy capacity region of a multiple-antenna gaussian broadcast channel with confidential messages. IEEE Trans Inf Theory 55(3):1235–1249
6. Tekin E, Yener A (2008) The Gaussian multiple access wire-tap channel. IEEE Trans Inf Theory 54(12):5747–5755
7. Liang Y, Poor HV (2008) Multiple-access channels with confidential messages. IEEE Trans Inf Theory 54(3):976–1002
8. Liang Y, Poor HV, Shamai (Shitz) S (2008) Information theoretic security. Found Trends Commun Inf Theory 5(4–5):355–580 (Now Publishers, Hanover)
9. Thangaraj A, Dihidar S, Calderbank AR, McLaughlin SW, Merolla J-M (2007) Applications of LDPC codes to the wiretap channel. IEEE Trans Inf Theory 53(8):2933–2945
10. Mahdavifar H, Vardy A (2011) Achieving the secrecy capacity of wiretap channels using polar codes. IEEE Trans Inf Theory 57(10):6428–6443
11. Klinc D, Ha J, McLaughlin SW, Barros J, Kwak B-J (2011) LDPC codes for the gaussian wiretap channel. IEEE Trans Inf Forensics Secur 6(3):532–540
12. Leung-Yan-Cheong SK, Hellman ME (1978) The Gaussian wire-tap channel. IEEE Trans Inf Theory IT–24(4):451–456
13. Khisti A, Wornell G (2010) Secure transmission with multiple antennas I: the MISOME wiretap channel. IEEE Trans Inf Theory 56(7):3088–3104
14. Khisti A, Wornell G (2010) Secure transmission with multiple antennas II: the MIMOME wiretap channel. IEEE Trans Inf Theory 56(11):5515–5532
15. Oggier F, Hassibi B (2011) The secrecy capacity of the MIMO wiretap channel. IEEE Trans Inf Theory 57(8):4961–4972
16. Bustin R, Liu R, Poor HV, Shamai (Shitz) S (2009) An MMSE approach to the secrecy capacity of the MIMO Gaussian wiretap channel. EURASIP J Wirel Commun Netw
17. Liang Y, Kramer G, Poor HV, Shamai (Shitz) S (2009) Compound wiretap channels. EURASIP J Wirel Commun Netw 2009:1–12
18. Liu T, Prabhakaran V, Vishwanath S (2008) The secrecy capacity of a class of parallel Gaussian compound wiretap channels. In: Proceedings of IEEE International Symposium on Information Theory (ISIT)
19. Liang Y, Poor HV, Shamai (Shitz) S (2008) Secure communication over fading channels. IEEE Trans Inf Theory 54(6):2470–2492
20. Gopala PK, Lai L, El Gamal H (2008) On the secrecy capacity of fading channels. IEEE Trans Inf Theory 54(10):4687–4698
21. Li J, Petropulu AP (2011) On ergodic secrecy rate for Gaussian MISO wiretap channels. IEEE Trans Wirel Commun 10(4):1176–1187
22. Lin S-C, Lin P-H (2013) On secrecy capacity of fast fading multiple-input wiretap channels with statistical CSIT. IEEE Trans Inf Forensics Secur 8(2):414–419
23. Bloch M, Barros J, Rodrigues MRD (2008) Wireless information-theoretic security. IEEE Trans Inf Theory 54(6):2515–2534
24. Prabhu VU, Rodrigues MRD (2011) On wireless channels with M-antenna eavesdroppers: characterization of the outage probability and ϵ-outage secrecy capacity. IEEE Trans Inf Forensics Secur 6(3):853–860
25. Chrysikos T, Dagiuklas T, Kotsopoulos S (2009) A closed-form expression for outage secrecy capacity in wireless information-theoretic security. In: Lecture Notes of the Institute for Computer Sciences. Social Informatics and Telecommunications Engineering
26. Gerbracht S, Scheunert C, Jorswieck EA (2012) Secrecy outage in MISO systems with partial channel information. IEEE Trans Inf Forensics Secur 7(2):704–716

Chapter 3
Secrecy Precoding and Beamforming in Multi-Antenna Wireless Systems

Abstract This chapter reviews various secrecy-enhancing signal processing techniques for data transmission in multi-antenna wireless systems. In particular, secrecy beamforming and precoding schemes are introduced as effective schemes to exploit the spatial degrees of freedom in multiple-input multiple-output (MIMO) systems. In these schemes, signals are directed towards spatial dimensions that yield large differences between the signal quality at the destination and that at the eaves-dropper. The use of artificial noise (AN) or jamming signals is also introduced as a way to further degrade the eavesdropper's reception (and, thus, more effectively enhance the signal quality difference between the two channels). The latter is especially useful when only partial knowledge of the eavesdropper's channel is available at the source.

Keywords Multiple-input multiple output (MIMO) · Beamforming · Precoding · Artificial noise · Secrecy

Motivated by results in the literature on information-theoretic secrecy, secrecy-enhancing signal processing techniques were proposed to construct channels that can result in large differences between the signal qualities at the destination and the eavesdropper. This can be achieved in both the data transmission and the channel estimation phases. In this chapter, we focus on signal designs for data transmission and leave discussions on channel estimation to Chap. 5. In the data transmission phase, signal quality difference can be achieved in two ways, namely, by using (i) secrecy beamforming and precoding schemes, with which signals are driven towards spatial dimensions that yield large differences between the signal qualities at the destination and the eavesdropper, and (ii) artificial noise (AN) transmissions, in which jamming signals are emitted to degrade the signal quality at the eavesdropper. Effective signal processing and coding schemes are both essential to achieve high secrecy rates, yet their designs are often closely coupled. In particular, well-designed signal processing techniques can reduce the complexity of wiretap code designs and, thus, help increase the achievable secrecy rate for more complicated scenarios, in which optimal coding schemes have not yet been determined. While this chapter focuses on

Y.-W. P. Hong et al., *Signal Processing Approaches to Secure Physical Layer Communications in Multi-Antenna Wireless Systems*, SpringerBriefs in Signal Processing, DOI: 10.1007/978-981-4560-14-6_3, © The Author(s) 2014

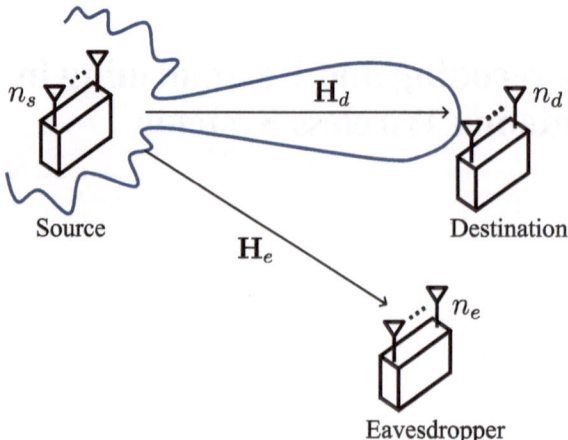

Fig. 3.1 An illustration of MIMO wiretap channel

the signal processing aspect of the physical-layer secrecy problem, one should keep in mind that the use of wiretap codes is inherently assumed when secrecy rate or secrecy capacity is utilized as the performance measure. The secrecy beamforming and precoding as well as AN transmission schemes are detailed in the following sections.

3.1 Secrecy Beamforming and Precoding Schemes for the Basic Multi-Antenna Wiretap Channel

In this section, secrecy beamforming and precoding schemes [1–4] are examined for the basic multi-antenna wiretap channel, which consists of only a single source, a single destination, and a single eavesdropper. With multiple antennas at the source, the information-bearing signal can be directed towards spatial dimensions that yield more favorable channel conditions toward the destination than the eavesdropper. This is referred to as secrecy beamforming when the signal is embedded within only a single dimension and is referred to as secrecy precoding when, more generally, multiple spatial dimensions are utilized. Secrecy beamforming can in be viewed as a special case of secrecy precoding but is discussed separately in the following since its design often yields useful insights.

3.1.1 Secrecy Beamforming for the Basic Multi-Antenna Wiretap Channel

Let us consider a multi-antenna wiretap channel that consists of a source, a desti-nation, and an eavesdropper as illustrated in Fig. 3.1. The nodes are equipped with n_s, n_d, and n_e antennas, respectively. Suppose that $\mathbf{x} \in \mathbb{C}^{n_s \times 1}$ is the signal trans-mitted by the source. Then, the signal received at the destination can be written as

$$\mathbf{y} = \mathbf{H}_d \mathbf{x} + \mathbf{w}, \tag{3.1a}$$

$$\mathbf{z} = \mathbf{H}_e \mathbf{x} + \mathbf{v}, \tag{3.1b}$$

where $\mathbf{H}_d \in \mathbb{C}^{n_d \times n_s}$ and $\mathbf{H}_e \in \mathbb{C}^{n_e \times n_s}$ are the source-to-destination and the source-to-eavesdropper channel matrices, and $\mathbf{w} \in \mathbb{C}^{n_d \times 1}$ and $\mathbf{v} \in \mathbb{C}^{n_e \times 1}$ are the addi-tive white Gaussian noise (AWGN) vectors at the destination and the eavesdropper, respectively. The AWGN vectors are assumed to have i.i.d. entries with zero-mean and unit-variance, i.e., $\mathbf{w} \sim \mathcal{CN}(0, \mathbf{I}_{n_d})$ and $\mathbf{v} \sim \mathcal{CN}(0, \mathbf{I}_{n_e})$.

Suppose that beamforming is adopted so that only a single-dimensional signal is transmitted by the source in each symbol period. In this case, the signal transmitted by the source can be expressed as

$$\mathbf{x} = \mathbf{f}u, \tag{3.2}$$

where $\mathbf{f} \in \mathbb{C}^{n_s \times 1}$ is the beamforming vector and u is the encoded symbol with zero mean and unit variance, i.e., $u \sim \mathcal{CN}(0, 1)$. The beamforming vector $\mathbf{f}$ must satisfy the transmit power constraint $\|\mathbf{f}\|^2 \leq \bar{P}$. By substituting (3.2) into (3.1), the received signals can be written as

$$\mathbf{y} = \mathbf{H}_d \mathbf{f}u + \mathbf{w}, \tag{3.3a}$$

$$\mathbf{z} = \mathbf{H}_e \mathbf{f}u + \mathbf{v}. \tag{3.3b}$$

This can be viewed as a special case of the MIMO wiretap channel with effective main and eavesdropper channel vectors given by $\mathbf{H}_d \mathbf{f}$ and $\mathbf{H}_e \mathbf{f}$, respectively. Let us express the beamforming vector as $\mathbf{f} = \sqrt{p_f}\tilde{\mathbf{f}}$, where $\tilde{\mathbf{f}} = \mathbf{f}/\|\mathbf{f}\|$ is the normalized beamforming vector and p_f is the power. Therefore, by Theorem 2.4, the maximum secrecy rate that is achievable with beamforming can be written as

$$R_{s,\mathrm{BF}} = \max_{\mathbf{f}:\|\mathbf{f}\|^2 \leq \bar{P}} \log \frac{\det(\mathbf{I}_{n_d} + \mathbf{H}_d \mathbf{f}\mathbf{f}^H \mathbf{H}_d^H)}{\det(\mathbf{I}_{n_e} + \mathbf{H}_e \mathbf{f}\mathbf{f}^H \mathbf{H}_e^H)} \tag{3.4}$$

$$= \max_{\tilde{\mathbf{f}}, p_f:\|\tilde{\mathbf{f}}\|^2=1, p_f \leq \bar{P}} \log \frac{1 + p_f \|\mathbf{H}_d \tilde{\mathbf{f}}\|^2}{1 + p_f \|\mathbf{H}_e \tilde{\mathbf{f}}\|^2} \tag{3.5}$$

which follows from Sylvester's determinant theorem [5].

Notice that the ratio inside the logarithm of (3.5) increases monotonically with p_f, when $\|\mathbf{H}_d\tilde{\mathbf{f}}\|^2 \geq \|\mathbf{H}_e\tilde{\mathbf{f}}\|^2$, and decreases monotonically, otherwise. Hence, the secrecy rate can be maximized by choosing $p_f = \bar{P}$ if there exists $\tilde{\mathbf{f}}$ such that $\|\mathbf{H}_d\tilde{\mathbf{f}}\|^2 \geq \|\mathbf{H}_e\tilde{\mathbf{f}}\|^2$ and by choosing $p_f = 0$, otherwise. As a result, the achievable secrecy rate can be expressed as

$$
R_{s,\mathrm{BF}} = \left[\max_{\tilde{\mathbf{f}}:\|\tilde{\mathbf{f}}\|^2=1} \log \frac{1 + \bar{P}\|\mathbf{H}_d\tilde{\mathbf{f}}\|^2}{1 + \bar{P}\|\mathbf{H}_e\tilde{\mathbf{f}}\|^2}\right]^+ = \left[\max_{\tilde{\mathbf{f}}:\|\tilde{\mathbf{f}}\|^2=1} \log \frac{\tilde{\mathbf{f}}^H(\mathbf{I}_{n_s} + \bar{P}\mathbf{H}_d^H\mathbf{H}_d)\tilde{\mathbf{f}}}{\tilde{\mathbf{f}}^H(\mathbf{I}_{n_s} + \bar{P}\mathbf{H}_e^H\mathbf{H}_e)\tilde{\mathbf{f}}}\right]^+.
$$

(3.6)

From the above, the optimal beamforming direction $\tilde{\mathbf{f}}$ can be found by maximizing the ratio inside the logarithm. Since $\mathbf{I}_{n_s} + \bar{P}\mathbf{H}_e^H\mathbf{H}_e$ is positive definite, there exists invertible matrix $\mathbf{D}$ such that $\mathbf{I}_{n_s} + \bar{P}\mathbf{H}_e^H\mathbf{H}_e = \mathbf{D}^H\mathbf{D}$. Then, by taking $\mathbf{g} \triangleq \mathbf{D}\tilde{\mathbf{f}}$, the problem reduces to finding $\mathbf{g}$ that maximizes

$$
\frac{\mathbf{g}^H\mathbf{D}^{-H}(\mathbf{I}_{n_s} + \bar{P}\mathbf{H}_d^H\mathbf{H}_d)\mathbf{D}^{-1}\mathbf{g}}{\mathbf{g}^H\mathbf{g}}.
$$

(3.7)

By Rayleigh-Ritz Theorem [6], the value of $\mathbf{g}$ that maximizes (3.7), denoted by $\mathbf{g}^*$, is the eigenvector corresponding to the maximum eigenvalue $\lambda_{\max}$ of the matrix $\mathbf{D}^{-H}(\mathbf{I}_{n_s} + \bar{P}\mathbf{H}_d^H\mathbf{H}_d)\mathbf{D}^{-1}$, that is,

$$
\mathbf{D}^{-H}(\mathbf{I}_{n_s} + \bar{P}\mathbf{H}_d^H\mathbf{H}_d)\mathbf{D}^{-1}\mathbf{g}^* = \lambda_{\max}\mathbf{g}^*.
$$

(3.8)

By letting $\tilde{\mathbf{f}}^* = \mathbf{D}^{-1}\mathbf{g}^*$ and by the fact that $\mathbf{D}^H\mathbf{D} = \mathbf{I}_{n_s} + P\mathbf{H}_e^H\mathbf{H}_e$, it follows from (3.8) that

$$
(\mathbf{I}_{n_s} + \bar{P}\mathbf{H}_e^H\mathbf{H}_e)^{-1}(\mathbf{I}_{n_s} + \bar{P}\mathbf{H}_d^H\mathbf{H}_d)\tilde{\mathbf{f}}^* = \lambda_{\max}\tilde{\mathbf{f}}^*.
$$

(3.9)

That is, the optimal beamforming direction should be chosen as the eigenvector corresponding to the maximum eigenvalue of the matrix $(\mathbf{I}_{n_s} + \bar{P}\mathbf{H}_e^H\mathbf{H}_e)^{-1}$ $(\mathbf{I}_{n_s} + \bar{P}\mathbf{H}_d^H\mathbf{H}_d)$ or, in other words, the generalized eigenvector corresponding to the maximum generalized eigenvalue of the matrix pair $(\mathbf{I}_{n_s} + \bar{P}\mathbf{H}_d^H\mathbf{H}_d, \mathbf{I}_{n_s} + \bar{P}\mathbf{H}_e^H\mathbf{H}_e)$ [6].

Theorem 3.1 *Under the power constraint $\bar{P}$, the maximum secrecy rate achievable with beamforming is given by*

$$
R_{s,\mathrm{BF}}(\bar{P}) = \left[\log \lambda_{\max}(\mathbf{I}_{n_s} + \bar{P}\mathbf{H}_d^H\mathbf{H}_d, \mathbf{I}_{n_s} + \bar{P}\mathbf{H}_e^H\mathbf{H}_e)\right]^+,
$$

(3.10)

where $\lambda_{\max}(\mathbf{A}, \mathbf{B})$ is the maximum generalized eigenvalue of the matrix pair $(\mathbf{A}, \mathbf{B})$ (i.e., the maximum eigenvalue of $\mathbf{B}^{-1}\mathbf{A}$ [7, 8]), and the corresponding optimal beamforming vector is given by [2]

$$\mathbf{f}^* = \sqrt{\bar{P}} \cdot \psi_{\max}\left(\mathbf{I}_{n_s} + \bar{P}\mathbf{H}_d^H\mathbf{H}_d, \mathbf{I}_{n_s} + \bar{P}\mathbf{H}_e^H\mathbf{H}_e\right), \qquad (3.11)$$

where $\psi_{\max}(\mathbf{A}, \mathbf{B})$ is the generalized eigenvector corresponding to the maximum generalized eigenvalue $\lambda_{\max}(\mathbf{A}, \mathbf{B})$.

Notice that, as the total transmit power $\bar{P}$ goes to infinity, the achievable secrecy rate becomes [2]

$$\lim_{\bar{P}\to\infty} R_{s,\mathrm{BF}}(\bar{P}) = \left[\log \lambda_{\max}(\mathbf{H}_d^H\mathbf{H}_d, \mathbf{H}_e^H\mathbf{H}_e)\right]^+. \qquad (3.12)$$

In particular, when the main channel $\mathbf{H}_d$ has components lying in the null space of the eavesdropper channel $\mathbf{H}_e$ (i.e., $(\mathbf{I} - \mathbf{H}_e(\mathbf{H}_e^H\mathbf{H}_e)^{-1}\mathbf{H}_e^H)\mathbf{H}_d^H \neq \mathbf{0}$), the maximum achievable secrecy rate approaches infinity since the source can steer its signal in a direction that nulls out the reception at the eavesdropper but not at the destination. This approach is referred to as *zero-forcing (ZF) beamforming*, which will also be utilized in later sections for more complicated scenarios. On the other hand, when $(\mathbf{I} - \mathbf{H}_e(\mathbf{H}_e^H\mathbf{H}_e)^{-1}\mathbf{H}_e^H)\mathbf{H}_d^H = \mathbf{0}$, the achievable secrecy rate would be bounded even as $\bar{P}$ goes to infinity since the increase in power benefits both the destination and the eavesdropper.

It is worthwhile to mention that beamforming is only a feasible solution within the general set of input covariance matrices considered in Theorem 2.4, i.e., $\{\mathbf{K_x} \triangleq \mathrm{E}[\mathbf{xx}^H] : \mathbf{K_x} \succeq \mathbf{0} \text{ and } \mathrm{tr}(\mathbf{K_x}) \leq \bar{P}\}$, where $\mathbf{K_x} \triangleq \mathrm{E}[\mathbf{xx}^H]$ is the covariance of the input signal $\mathbf{x}$. In particular, it can be considered as the case where $\mathbf{K_x} = \mathbf{ff}^H$ and was shown to be optimal (i.e., secrecy-capacity achieving) in [2] for the case where only a single antenna exists at the destination (i.e., the case with $n_d = 1$) and in [9] for the case where $\mathbf{H}_d^H\mathbf{H}_d - \mathbf{H}_e^H\mathbf{H}_e$ yields only one positive eigenvalue. That is, for the case where only a single antenna exists at the destination, the secrecy capacity is given by

$$C_s(\bar{P}) = \left[\log \lambda_{\max}(\mathbf{I}_{n_s} + \bar{P}\mathbf{h}_d^H\mathbf{h}_d, \mathbf{I}_{n_s} + \bar{P}\mathbf{H}_e^H\mathbf{H}_e)\right]^+, \qquad (3.13)$$

where $\mathbf{h}_d \in \mathbb{C}^{1\times n_s}$ is a row vector representing a special case of the channel matrix between the source and the destination for $n_d = 1$. These results are rather intuitive since, in the first case, the destination can receive only over a single dimension and, thus, no advantage can be gained by transmitting over multiple dimensions. In the second case, only a single dimension yields a more favorable channel at the destination than the eavesdropper and, thus, a positive secrecy rate can only be achieved in this direction.

3.1.2 Secrecy Precoding for the Basic Multi-Antenna Wiretap Channel

When the destination is equipped with multiple antennas, confidential messages can be spatially multiplexed onto multiple independent subchannels via precoding. By Theorem 2.4, the secrecy capacity in this case is achieved by choosing the channel input $\mathbf{x}$ to be Gaussian and can be expressed as

$$C_s = \max_{\mathbf{K_x} \succeq 0, \mathrm{tr}(\mathbf{K_x}) \leq \bar{P}} \log \frac{\det(\mathbf{I}_{n_d} + \mathbf{H}_d \mathbf{K_x} \mathbf{H}_d^H)}{\det(\mathbf{I}_{n_e} + \mathbf{H}_e \mathbf{K_x} \mathbf{H}_e^H)}. \tag{3.14}$$

Notice that the secrecy capacity expression involves an optimization over the input covariance matrix $\mathbf{K_x}$. For any $\mathbf{K_x} \succeq \mathbf{0}$ with rank$(\mathbf{K_x}) = k_s$, there exists $\mathbf{F} \in \mathbb{C}^{n_s \times k_s}$ such that the input vector can be expressed as $\mathbf{x} = \mathbf{Fu}$, where $\mathbf{u} \sim \mathcal{CN}(\mathbf{0}, \mathbf{I}_{k_s})$ correspond to symbols in a vector Gaussian codebook. Here, $\mathbf{F}$ can be viewed as the secrecy precoding matrix and $\mathbf{u}$ can be viewed as the vector with elements associated with k_s encoded data streams. Therefore, the search for the optimal secrecy precoding matrix can be obtained via the search for $\mathbf{K_x}$. However, the problem in (3.14) is non-convex over the set of covariance matrices that satisfy the total power constraint, i.e., $\{\mathbf{K_x} : \mathbf{K_x} \succeq \mathbf{0} \text{ and } \mathrm{tr}(\mathbf{K_x}) \leq \bar{P}\}$, and, thus, is difficult to solve in general.

Instead, one can consider the power-covariance constraint, as proposed in [10], where the input covariance matrix is constrained by a positive definite power-covariance matrix $\mathbf{S}$, i.e., $\mathbf{0} \preceq \mathbf{K_x} \preceq \mathbf{S}$. Notice that the power-covariance constraint limits both the transmit power on each individual antenna and the correlation between their respective channel inputs. Thus, it is stricter than the total power constraint. Suppose that $C_s(\mathbf{S})$ is the secrecy capacity under the power-covariance constraint given by $\mathbf{0} \preceq \mathbf{K_x} \preceq \mathbf{S}$. Then, the secrecy capacity under the total power constraint can be obtained by maximizing over all matrices $\mathbf{S}$ that satisfy the total power constraint $\bar{P}$, i.e.,

$$C_s = \max_{\mathbf{S}:\mathbf{S} \succeq 0, \mathrm{tr}(\mathbf{S}) \leq \bar{P}} C_s(\mathbf{S}). \tag{3.15}$$

For simplicity, let us first consider the case where $n_s = n_d = n_e$ and where $\mathbf{H}_d$ and $\mathbf{H}_e$ are assumed to be invertible. In this case, the received signal vectors at the destination and the eavesdropper can be multipled by $\mathbf{H}_d^{-1}$ and $\mathbf{H}_e^{-1}$, respectively, and the effective channel model can be written equivalently as

$$\tilde{\mathbf{y}} = \mathbf{H}_d^{-1} \mathbf{y} = \mathbf{x} + \tilde{\mathbf{w}}, \tag{3.16a}$$

$$\tilde{\mathbf{z}} = \mathbf{H}_e^{-1} \mathbf{z} = \mathbf{x} + \tilde{\mathbf{v}}, \tag{3.16b}$$

where $\tilde{\mathbf{w}} = \mathbf{H}_d^{-1} \mathbf{w}$ and $\tilde{\mathbf{v}} = \mathbf{H}_d^{-1} \mathbf{v}$ are the effective AWGN vectors at the destination and the eavesdropper with $\mathbf{K}_{\tilde{\mathbf{w}}} \triangleq \mathrm{E}[\tilde{\mathbf{w}} \tilde{\mathbf{w}}^H] = \mathbf{H}_d^{-1} \mathbf{H}_d^{-H}$ and $\mathbf{K}_{\tilde{\mathbf{v}}} \triangleq \mathrm{E}[\tilde{\mathbf{v}} \tilde{\mathbf{v}}^H] = \mathbf{H}_e^{-1} \mathbf{H}_e^{-H}$, respectively.

Following the approach in [10], let $\mathbf{S}^{\frac{1}{2}}$ be an $n_s \times n_s$ invertible matrix chosen such that

$$\mathbf{S} = \mathbf{S}^{\frac{1}{2}}(\mathbf{S}^{\frac{1}{2}})^H \tag{3.17}$$

and let $\mathbf{C}$ be an invertible generalized eigenvector matrix of the two symmetric positive definite matrices $\mathbf{I}_{n_s} + (\mathbf{S}^{\frac{1}{2}})^H \mathbf{K}_{\tilde{\mathbf{v}}}^{-1} \mathbf{S}^{\frac{1}{2}}$ and $\mathbf{I}_{n_s} + (\mathbf{S}^{\frac{1}{2}})^H \mathbf{K}_{\tilde{\mathbf{w}}}^{-1} \mathbf{S}^{\frac{1}{2}}$ such that

$$\mathbf{C}^H \left(\mathbf{I}_{n_s} + (\mathbf{S}^{\frac{1}{2}})^H \mathbf{K}_{\tilde{\mathbf{v}}}^{-1} \mathbf{S}^{\frac{1}{2}} \right) \mathbf{C} = \mathbf{I}_{n_s} \tag{3.18}$$

and

$$\mathbf{C}^H \left(\mathbf{I}_{n_s} + (\mathbf{S}^{\frac{1}{2}})^H \mathbf{K}_{\tilde{\mathbf{w}}}^{-1} \mathbf{S}^{\frac{1}{2}} \right) \mathbf{C} = \mathbf{\Lambda}_d, \tag{3.19}$$

where $\mathbf{\Lambda}_d = \mathrm{diag}\left(\lambda_1, \ldots, \lambda_{n_s} \right)$ is a positive definite diagonal matrix. Let b be the number of diagonal elements in $\mathbf{\Lambda}_d$ that are greater than 1, i.e., $\lambda_1 \geq \cdots \geq \lambda_b \geq 1 \geq \cdots \geq \lambda_{n_s}$. The diagonal matrix can then be expressed in terms of the two submatrices $\mathbf{\Lambda}_{d1} = \mathrm{diag}(\lambda_1, \ldots, \lambda_b)$ and $\mathbf{\Lambda}_{d2} = \mathrm{diag}(\lambda_{b+1}, \ldots, \lambda_{n_s})$ such that

$$\mathbf{\Lambda}_d = \begin{bmatrix} \mathbf{\Lambda}_{d1} & \mathbf{0} \\ \mathbf{0} & \mathbf{\Lambda}_{d2} \end{bmatrix}.$$

By extending the results in [10] to the complex case, the optimal input covariance matrix can be expressed as

$$\mathbf{K}_{\mathbf{x}}^* = \mathbf{F}\mathbf{F}^H = \mathbf{S}^{\frac{1}{2}}\mathbf{C} \left(\begin{matrix} (\mathbf{C}_1^H \mathbf{C}_1)^{-1} & \mathbf{0} \\ \mathbf{0} & \mathbf{0} \end{matrix} \right) \mathbf{C}^H (\mathbf{S}^{\frac{1}{2}})^H, \tag{3.20}$$

where $\mathbf{C} = [\mathbf{C}_1\ \mathbf{C}_2]$ with $\mathbf{C}_1$ being an $n_s \times b$ submatrix. It can be easily verified that $\mathbf{K}_{\mathbf{x}}^*$ indeed satisfies the power covariance constraint $\mathbf{0} \preceq \mathbf{K}_{\mathbf{x}}^* \preceq \mathbf{S}$. Given that the channel input is Gaussian with covariance $\mathbf{K}_{\mathbf{x}}^*$, the achievable secrecy rate is given by

$$R_s(\mathbf{S}) = \log \frac{\det(\mathbf{I}_{n_d} + \mathbf{H}_d \mathbf{K}_{\mathbf{x}}^* \mathbf{H}_d^H)}{\det(\mathbf{I}_{n_e} + \mathbf{H}_e \mathbf{K}_{\mathbf{x}}^* \mathbf{H}_e^H)} = \log \frac{\det(\mathbf{I}_{n_s} + \mathbf{K}_{\mathbf{x}}^* \mathbf{H}_d^H \mathbf{H}_d)}{\det(\mathbf{I}_{n_s} + \mathbf{K}_{\mathbf{x}}^* \mathbf{H}_e^H \mathbf{H}_e)}. \tag{3.21}$$

By reorganizing (3.19) and by substituting $\mathbf{K}_{\tilde{\mathbf{w}}}$ with $\mathbf{H}_d^{-1} \mathbf{H}_d^{-H}$, we have

$$\mathbf{H}_d^H \mathbf{H}_d = (\mathbf{S}^{\frac{-1}{2}})^H \left[\mathbf{C}^{-H} \left(\begin{matrix} \mathbf{\Lambda}_{d1} & \mathbf{0} \\ \mathbf{0} & \mathbf{\Lambda}_{d2} \end{matrix} \right) \mathbf{C}^{-1} - \mathbf{I}_{n_s} \right] \mathbf{S}^{\frac{-1}{2}}. \tag{3.22}$$

It then follows from (3.20) and (3.22) that

$$\det(\mathbf{I}_{n_s} + \mathbf{K}_x^* \mathbf{H}_d^H \mathbf{H}_d) = \det\left(\begin{pmatrix} (\mathbf{C}_1^H \mathbf{C}_1)^{-1} \mathbf{\Lambda}_{d1} & -(\mathbf{C}_1^H \mathbf{C}_1)^{-1} \mathbf{C}_1^H \mathbf{C}_2 \\ \mathbf{0} & \mathbf{I} \end{pmatrix}\right) \quad (3.23)$$

$$= \det\left(\left(\mathbf{C}_1^H \mathbf{C}_1\right)^{-1}\right) \det(\mathbf{\Lambda}_{d1}). \quad (3.24)$$

Similarly, it can be shown that $\det(\mathbf{I}_{n_s} + \mathbf{K}_x^* \mathbf{H}_e^H \mathbf{H}_e) = \det\left(\left(\mathbf{C}_1^H \mathbf{C}_1\right)^{-1}\right) \det(\mathbf{I})$ and, thus, the secrecy rate in (3.21) is equal to

$$R_s(\mathbf{S}) = \log \det(\mathbf{\Lambda}_{d1}). \quad (3.25)$$

In [10], it was shown that the secrecy capacity of the MIMO wiretap channel under the power covariance constraint $\mathbf{S}$ can be upper-bounded as

$$C_s(\mathbf{S}) \leq \max_{0 \preceq \mathbf{K}_x \preceq \mathbf{S}} \log \frac{\det(\mathbf{I} + \mathbf{K}_x \mathbf{K}_0^{-1})}{\det(\mathbf{I} + \mathbf{K}_x \mathbf{K}_{\tilde{v}}^{-1})} = \log \frac{\det(\mathbf{I} + \mathbf{S}\mathbf{K}_0^{-1})}{\det(\mathbf{I} + \mathbf{S}\mathbf{K}_{\tilde{v}}^{-1})} \triangleq \bar{C}_s(\mathbf{S}) \quad (3.26)$$

where

$$\mathbf{K}_0 = \mathbf{S}^{\frac{1}{2}} \left[\mathbf{C}^{-H} \begin{pmatrix} \mathbf{\Lambda}_{d1} & \mathbf{0} \\ \mathbf{0} & \mathbf{I}_{n_s-b} \end{pmatrix} \mathbf{C}^{-1} - \mathbf{I}\right]^{-1} (\mathbf{S}^{\frac{1}{2}})^H. \quad (3.27)$$

It can be easily verified that $\mathbf{K}_0 \preceq \mathbf{K}_{\tilde{v}}$ and $\mathbf{K}_0 \preceq \mathbf{K}_{\tilde{w}}$. Therefore, (3.26) can be viewed as the secrecy capacity of a degraded MIMO wiretap channel with noise covariance matrices $\mathbf{K}_0$ and $\mathbf{K}_{\tilde{v}}$ at the destination and the eavesdropper, respectively. Interestingly, the upper bound in (3.26) can be further evaluated as

$$\bar{C}_s(\mathbf{S}) = \log \frac{\det(\mathbf{I} + (\mathbf{S}^{\frac{1}{2}})^H \mathbf{K}_0^{-1} \mathbf{S}^{\frac{1}{2}})}{\det(\mathbf{I} + (\mathbf{S}^{\frac{1}{2}})^H \mathbf{K}_e^{-1} \mathbf{S}^{\frac{1}{2}})} \quad (3.28)$$

$$= \log \frac{\det \mathbf{C}^H (\mathbf{I} + (\mathbf{S}^{\frac{1}{2}})^H \mathbf{K}_0^{-1} \mathbf{S}^{\frac{1}{2}}) \mathbf{C}}{\det \mathbf{C}^H (\mathbf{I} + (\mathbf{S}^{\frac{1}{2}})^H \mathbf{K}_e^{-1} \mathbf{S}^{\frac{1}{2}}) \mathbf{C}} \quad (3.29)$$

$$= \det(\mathbf{\Lambda}_{d1}). \quad (3.30)$$

That is, the secrecy rate achieved with the input covariance matrix $\mathbf{K}_x^*$, i.e., the rate given in (3.25), is equal to the secrecy capacity upper bound evaluated in (3.30). This shows that the secrecy capacity upper bound in (3.30) is tight and that the input covariance matrix $\mathbf{K}_x^*$ given in (3.20) is optimal.

Based on the solution given in (3.20), the optimal precoding matrix can be obtained as

$$\mathbf{F} = \mathbf{S}^{\frac{1}{2}} \mathbf{C} \begin{pmatrix} \mathbf{D} \\ \mathbf{0} \end{pmatrix} = \mathbf{S}^{\frac{1}{2}} \mathbf{C}_1 \mathbf{D} \quad (3.31)$$

where $\mathbf{D}$ is a $b \times b$ matrix chosen such that $(\mathbf{C}_1^H \mathbf{C}_1)^{-1} = \mathbf{D}\mathbf{D}^H$. The optimal precoding matrix given above shows that the confidential message should be trans-

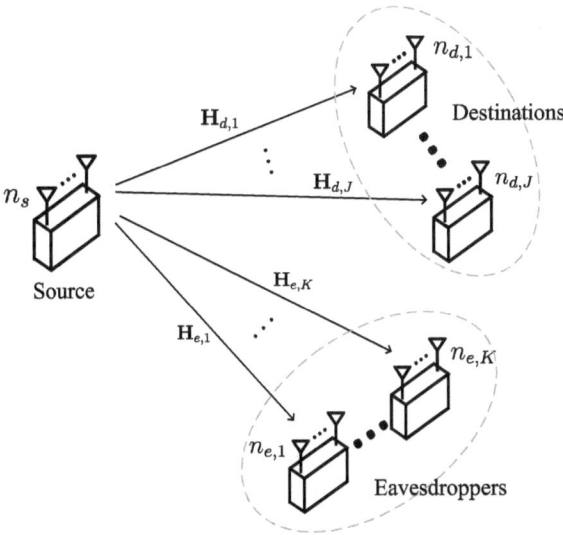

Fig. 3.2 An illustration of a multi-antenna wireless system with multiple destinations and multiple eavesdroppers

mitted only over dimensions associated with the generalized eigenvalues $\lambda_1, \ldots, \lambda_b$, namely, dimensions that yield better effective channels at the destination than at the eavesdropper.

3.2 Secrecy Beamforming for Systems with Multiple Destinations and Eavesdroppers

Secrecy beamforming and precoding techniques can also be applied to wiretap channels with multiple destinations and eavesdroppers, as illustrated in Fig. 3.2. For systems with multiple destinations, we can consider two scenarios: (i) the multicast scenario where a common confidential message is transmitted to all destinations simultaneously and (ii) the broadcast scenario where different confidential messages are intended for different destinations. The multicast scenario corresponds to the compound wiretap channel examined in [11] and in Chap. 2. Note that the secrecy capacity of the above-mentioned scenarios are generally unknown and the optimal secrecy beamforming and precoding schemes are not easy to obtain due to the non-convex structure of the problems. In the following, some tractable solutions are given for special cases, namely, those considered in [12–15].

3.2.1 Multicast Secrecy Beamforming

In this section, we consider secrecy beamforming for multicast applications where a common confidential message is to be transmitted to multiple destinations simultaneously. Let us consider a system that consists of a source with n_s antennas, J single-antenna destinations, and K multi-antenna eavesdroppers. The number of antennas at the kth eavesdropper is denoted by $n_{e,k}$. Suppose that $\mathbf{x}$ is the signal transmitted by the source. Then, the received signals at the jth destination and the kth eavesdropper can be written as

$$y_j = \mathbf{h}_{d,j}\mathbf{x} + w_k \tag{3.32}$$

$$\mathbf{z}_k = \mathbf{H}_{e,k}\mathbf{x} + \mathbf{v}_k \tag{3.33}$$

where $\mathbf{h}_{d,j} \in \mathbb{C}^{1 \times n_s}$ is the channel vector associated with the jth destination and $\mathbf{H}_{e,k} \in \mathbb{C}^{n_{e,k} \times n_s}$ is the channel matrix associated with the kth eavesdropper. The AWGN w_j and $\mathbf{v}_k$ are assumed to have entries that are i.i.d. Gaussian with mean 0 and variance 1. The case with multiple antennas at the destinations can be derived similarly.

Suppose that the secrecy beamforming scheme is employed and, thus, the input signal vector $\mathbf{x}$ is assumed to have a covariance matrix $\mathbf{K_x} \triangleq \mathrm{E}[\mathbf{x}\mathbf{x}^H]$ that has rank equal to 1, i.e., $\mathrm{rank}(\mathbf{K_x}) = 1$. In this case, the transmitted signal can be expressed as $\mathbf{x} = \mathbf{f}u$, where $\mathbf{f}$ is the $n_s \times 1$ beamforming vector with $\|\mathbf{f}\|^2 \leq \bar{P}$ and u is the encoded symbol with $\mathrm{E}[|u|^2] = 1$. Then, following (2.29), an achievable secrecy rate for this system can be given as

$$R_s = \max_{\mathbf{f}:\|\mathbf{f}\|^2 \leq \bar{P}} \min_{j,k} \; \log \frac{1 + |\mathbf{h}_{d,j}\mathbf{f}|^2}{1 + \|\mathbf{H}_{e,k}\mathbf{f}\|^2}, \tag{3.34}$$

$$= \max_{\substack{\mathbf{K_x} \succeq 0, \mathrm{tr}(\mathbf{K_x}) \leq \bar{P} \\ \mathrm{rank}(\mathbf{K_x})=1}} \min_{j,k} \; \log \frac{1 + \mathbf{h}_{d,j}\mathbf{K_x}\mathbf{h}_{d,j}^H}{1 + \mathrm{tr}(\mathbf{H}_{e,k}\mathbf{K_x}\mathbf{H}_{e,k}^H)}, \tag{3.35}$$

which is obtained by choosing u to be Gaussian. Note that the secrecy-rate maximization problem given in (3.35) is difficult to solve due to the rank-1 constraint. To obtain a solution to this problem, let us first consider a relaxation where the rank constraint is first dropped. By doing so, the problem becomes a quasi-convex problem whose globally optimal solution can be found using standard techniques such as the bisection algorithm [16]. Alternatively, one can also adopt the approach proposed in [13] where the relaxed optimization problem is reformulated as a semi-definite programming (SDP) problem [17] using the Charnes-Cooper transformation [18]. In this case, the optimality of the relaxed solution can be shown for certain cases as described below and in [13].

Specifically, by relaxing the rank constraint, the optimization problem in (3.35) can be reformulated as

$$\min_{\mathbf{K_x}} \frac{1 + \max_k \text{tr}(\mathbf{H}_{e,k} \mathbf{K_x} \mathbf{H}_{e,k}^H)}{1 + \min_j \mathbf{h}_{d,j} \mathbf{K_x} \mathbf{h}_{d,j}^H} \tag{3.36a}$$

$$\text{subject to } \mathbf{K_x} \succeq \mathbf{0}, \ \text{tr}(\mathbf{K_x}) \leq \bar{P}. \tag{3.36b}$$

By performing the change of variable $\mathbf{K_x} = \mathbf{Q}/\xi$ and $\xi > 0$, the problem can be equivalently formulated as

$$\min_{\mathbf{Q},\xi} \frac{\xi + \max_k \text{tr}(\mathbf{H}_{e,k} \mathbf{Q} \mathbf{H}_{e,k}^H)}{\xi + \min_j \mathbf{h}_{d,j} \mathbf{Q} \mathbf{h}_{d,j}^H} \tag{3.37a}$$

$$\text{subject to } \text{tr}(\mathbf{Q}) \leq \xi \bar{P}, \ \mathbf{Q} \succeq \mathbf{0}, \ \xi > 0 \tag{3.37b}$$

From the above, the problem can then be transformed into an equivalent SDP problem [13] as given below:

$$\min_{\mathbf{Q},\xi,\tau} \tau \tag{3.38a}$$

$$\text{subject to } \xi + \text{tr}(\mathbf{H}_{e,k} \mathbf{Q} \mathbf{H}_{e,k}^H) \leq \tau, \ \forall k \tag{3.38b}$$

$$\xi + \mathbf{h}_{d,j} \mathbf{Q} \mathbf{h}_{d,j}^H \geq 1, \ \forall j \tag{3.38c}$$

$$\text{tr}(\mathbf{Q}) \leq \xi \bar{P}, \ \mathbf{Q} \succeq \mathbf{0}, \ \xi \geq 0. \tag{3.38d}$$

The SDP problem can then be solved efficiently and reliably using off-the-shelf solvers such as SeDuMi [19] and CVX [20]. Details on the proof of equivalence can be found in the appendix of [13].

If the solution of the relaxed problem in (3.38), denoted by $\tilde{\mathbf{K}}_\mathbf{x}^*$, has rank equal to 1, then we are done since one can then choose beamforming vector $\mathbf{f}^*$ such that $\mathbf{f}^*(\mathbf{f}^*)^H = \tilde{\mathbf{K}}_\mathbf{x}^*$. However, due to the relaxation, the solution $\tilde{\mathbf{K}}_\mathbf{x}^*$ is not guaranteed to have rank equal to 1, in which case, extra procedures must be performed to extract the effective beamforming vectors from the solution $\tilde{\mathbf{K}}_\mathbf{x}^*$. There are several ways to do this. In particular, one can take the deterministic approach, where $\mathbf{f}^*$ is chosen as the principle eigenvector of $\tilde{\mathbf{K}}_\mathbf{x}^*$, or adopt the randomized approach, where a set of vectors $\{\mathbf{f}_1, \ldots, \mathbf{f}_M\}$ are chosen independently according to the distribution $\mathcal{CN}(\mathbf{0}, \tilde{\mathbf{K}}_\mathbf{x}^*)$ and the beamformer $\mathbf{f}^*$ is chosen as the normalized version of the vector that yields the largest secrecy rate. Readers are referred to [17] for further discussions on semidefinite relaxation methods.

It is worthwhile to remark that, in general, the use of beamforming, i.e., the use of a rank-1 covariance matrix, yields a secrecy rate that is less than that considered in (2.29). That is, secrecy precoding where data is multiplexed onto multiple dimensions may in fact perform better in certain scenarios. This is especially the case when multiple antennas exist at the destination as well. However, the optimal secrecy precoders for these scenarios are difficult to find in general. Interestingly, it was shown in [13] that secrecy beamforming is optimal for certain special cases, e.g., for systems with only $J = 1$ single-antenna destination (regardless of the number of eavesdroppers) and also for systems with $J \leq 3$ destinations and $K = 1$ eaves-dropper. The relaxed SDP problem in (3.38) then yields the optimal solution in these cases. In other cases, beamforming may still be desirable since it requires a much smaller coding complexity compared to multi-dimensional transmissions. Similar observations were made in the literature on multicast beamforming for non-secrecy applications, e.g., in [21].

In addition to the secrecy-rate maximization problem given in (3.35), secrecy beamformers can also be designed by considering an alternative optimization problem where the goal is to minimize the transmit power subject to constraints on the secrecy rate, i.e.,

$$\min_{\mathbf{K_x}} \text{tr}(\mathbf{K_x}) \tag{3.39a}$$

$$\text{subject to} \min_{j,k} \ \log \frac{1 + \mathbf{h}_{d,j} \mathbf{K_x} \mathbf{h}_{d,j}^H}{1 + \text{tr}(\mathbf{H}_{e,k} \mathbf{K_x} \mathbf{H}_{e,k}^H)} \geq R_0 \tag{3.39b}$$

$$\mathbf{K_x} \succeq \mathbf{0}, \ \text{rank}(\mathbf{K_x}) = 1. \tag{3.39c}$$

This problem can be reformulated as

$$\min_{\mathbf{K_x}} \text{tr}(\mathbf{K_x}) \tag{3.40a}$$

$$\text{subject to } 1 + \mathbf{h}_{d,j} \mathbf{K_x} \mathbf{h}_{d,j}^H \geq 2^{R_0}[1 + \text{tr}(\mathbf{H}_{e,k} \mathbf{K_x} \mathbf{H}_{e,k}^H)], \ \forall j, k \tag{3.40b}$$

$$\mathbf{K_x} \succeq \mathbf{0}, \ \text{rank}(\mathbf{K_x}) = 1. \tag{3.40c}$$

Again, this problem is difficult to solve due to the non-convex constraint rank$(\mathbf{K_x}) = 1$. In this case, we can consider a similar approach as before where we first drop the rank-1 constraint and then extract an approximate solution from the relaxed SDP problem [17].

3.2.2 Broadcast Secrecy Beamforming

In this section, we consider the broadcasting application where different data-streams are to be transmitted simultaneously to multiple destinations. In this case, a different beamformer must be associated with each data-stream and the design must take into consideration interference from other data streams. This problem was considered in the context of secure satellite communications in, e.g., [14, 15], and is related to the study of broadcast channels with confidential messages in the information-theoretic literature [22–24].

Let us consider a system that consists of a source with n_s antennas, J single-antenna destinations, and K single-antenna eavesdroppers. Let u_j be the data intended for destination j with $\text{E}[|u_j|^2] = 1$ and let $\mathbf{f}_j$ be the beamforming vector used to transmit u_j. The signal transmitted by the source is a summation of the signals intended for all destinations, i.e.,

$$\mathbf{x} = \sum_{j=1}^{J} \mathbf{f}_j u_j. \tag{3.41}$$

The signal received at destination j and eavesdropper k are given by

$$y_j = \mathbf{h}_{d,j} \mathbf{f}_j u_j + \mathbf{h}_{d,j} \sum_{\ell \neq j} \mathbf{f}_\ell u_\ell + w_j \tag{3.42}$$

and

$$z_k = \mathbf{h}_{e,k} \mathbf{f}_j u_j + \mathbf{h}_{e,k} \sum_{\ell \neq j} \mathbf{f}_\ell u_\ell + v_k, \tag{3.43}$$

respectively, where $\mathbf{h}_{d,j} \in \mathbb{C}^{1 \times n_s}$ and $\mathbf{h}_{e,k} \in \mathbb{C}^{1 \times n_s}$ are the channels to destination j and eavesdropper k, respectively, and w_j, $v_k \sim \mathcal{CN}(0, 1)$ are the AWGN. By (2.29), an achievable secrecy rate at destination j can be given as

$$R_{s,j} = \log \frac{1 + \dfrac{|\mathbf{h}_{d,j} \mathbf{f}_j|^2}{1 + \sum_{\ell \neq j} |\mathbf{h}_{d,j} \mathbf{f}_\ell|^2}}{1 + \max_k \dfrac{|\mathbf{h}_{e,k} \mathbf{f}_j|^2}{1 + \sum_{\ell \neq j} |\mathbf{h}_{e,k} \mathbf{f}_\ell|^2}} = \log \frac{1 + \dfrac{\mathbf{h}_{d,j} \mathbf{K}_j \mathbf{h}_{d,j}^H}{1 + \sum_{\ell \neq j} \mathbf{h}_{d,j} \mathbf{K}_\ell \mathbf{h}_{d,j}^H}}{1 + \max_k \dfrac{\mathbf{h}_{e,k} \mathbf{K}_j \mathbf{h}_{e,k}^H}{1 + \sum_{\ell \neq j} \mathbf{h}_{e,k} \mathbf{K}_\ell \mathbf{h}_{e,k}^H}} \tag{3.44}$$

where $\mathbf{K}_j = \mathbf{f}_j \mathbf{f}_j^H$ is the rank-1 covariance matrix corresponding to the signal intended for destination j, i.e., $\mathbf{f}_j u_j$.

To obtain the secrecy beamformer, let us consider the secrecy-rate-constrained power-minimization problem proposed in [14, 15], i.e.,

$$\min_{\mathbf{K}_j, \forall j} \sum_{j=1}^{J} \mathrm{tr}(\mathbf{K}_j) \tag{3.45a}$$

$$\text{subject to} \quad \frac{1 + \dfrac{\mathbf{h}_{d,j} \mathbf{K}_j \mathbf{h}_{d,j}^H}{1 + \sum_{\ell \neq j} \mathbf{h}_{d,j} \mathbf{K}_\ell \mathbf{h}_{d,j}^H}}{1 + \max_k \dfrac{\mathbf{h}_{e,k} \mathbf{K}_j \mathbf{h}_{e,k}^H}{1 + \sum_{\ell \neq j} \mathbf{h}_{e,k} \mathbf{K}_\ell \mathbf{h}_{e,k}^H}} \geq 2^{R_{0,j}} \tag{3.45b}$$

$$\mathbf{K}_j \succeq \mathbf{0}, \ \mathrm{rank}(\mathbf{K}_j) = 1, \ \forall j. \tag{3.45c}$$

By introducing auxiliary variables $\alpha_1, \ldots, \alpha_J$, the problem can be equivalently formulated as

$$\min_{\mathbf{K}_j, \alpha_j, \forall j} \sum_{j=1}^{J} \mathrm{tr}(\mathbf{K}_j) \tag{3.46a}$$

$$\text{subject to } 1 + \frac{\mathbf{h}_{d,j} \mathbf{K}_j \mathbf{h}_{d,j}^H}{1 + \sum_{\ell \neq j} \mathbf{h}_{d,j} \mathbf{K}_\ell \mathbf{h}_{d,j}^H} \geq \alpha_j 2^{R_{0,j}} \tag{3.46b}$$

$$1 + \max_{k} \frac{\mathbf{h}_{e,k} \mathbf{K}_j \mathbf{h}_{e,k}^H}{1 + \sum_{\ell \neq j} \mathbf{h}_{e,k} \mathbf{K}_\ell \mathbf{h}_{e,k}^H} \leq \alpha_j \tag{3.46c}$$

$$\mathbf{K}_j \succeq \mathbf{0}, \ \mathrm{rank}(\mathbf{K}_j) = 1, \ \alpha_j \geq 0, \ \forall j, \tag{3.46d}$$

which can be further reorganized as

$$\min_{\mathbf{K}_j, \alpha_j, \forall j} \sum_{j=1}^{J} \mathrm{tr}(\mathbf{K}_j) \tag{3.47a}$$

$$\text{subject to } \mathbf{h}_{d,j} \mathbf{K}_j \mathbf{h}_{d,j}^H - (\alpha_j 2^{R_{0,j}} - 1) \sum_{\ell \neq j} \mathbf{h}_{d,j} \mathbf{K}_\ell \mathbf{h}_{d,j}^H \geq \alpha_j 2^{R_{0,j}} - 1, \tag{3.47b}$$

$$\mathbf{h}_{e,k} \mathbf{K}_j \mathbf{h}_{e,k}^H - (\alpha_j - 1) \sum_{\ell \neq j} \mathbf{h}_{e,k} \mathbf{K}_\ell \mathbf{h}_{e,k}^H \leq \alpha_j - 1, \ \forall k, \tag{3.47c}$$

$$\mathbf{K}_j \succeq \mathbf{0}, \ \mathrm{rank}(\mathbf{K}_j) = 1, \ \alpha_j \geq 0, \ \forall j. \tag{3.47d}$$

Notice that this problem is still difficult to solve since it is non-convex. However, an efficient solution can be obtained, as proposed in [15], by considering the semi-definite relaxation approach where the rank-1 constraint is first relaxed. By relaxing the rank-1 constraint and by fixing the variables $\alpha_1, \ldots, \alpha_J$, the problem becomes a standard SDP problem and can be efficiently solved using off-the-shelf solvers, e.g., SeDuMi [19] and CVX [20]. The optimized objective (i.e., the sum power) for fixed $\boldsymbol{\alpha} = (\alpha_1, \ldots, \alpha_J)$, denoted by $\mathrm{tr}(\mathbf{K_x}(\boldsymbol{\alpha}))$, was shown in [15] to be convex with respect to $\boldsymbol{\alpha}$. Therefore, the optimization over $\boldsymbol{\alpha}$ can then be performed using efficient gradient-based methods [15]. It is necessary to remark that the solution obtained via this procedure may not be rank-1 and, thus, randomization techniques [17] must be employed to extract from it the desired rank-1 solution.

It is interesting to remark that the procedure given above is based on numerical methods and does not provide a closed-form expression of the secrecy beamformer. If a closed-form expression is desired, one can employ the simpler but less effective zero-forcing (ZF) based secrecy beamforming schemes, where the beamformer is chosen to null out the signals received by all eavesdroppers as well as the cochannel interference to all other destinations [14, 15]. That is, the beamformer for destination j, i.e., $\mathbf{f}_j$, is chosen such that $\mathbf{h}_{e,k} \mathbf{f}_j = 0$ and $\mathbf{h}_{d,\ell} \mathbf{f}_j = 0$ for all k and for all $\ell \neq j$. Let $\tilde{\mathbf{H}}_j = \left[\mathbf{h}_{d,1}^T, \ldots, \mathbf{h}_{d,j-1}^T, \mathbf{h}_{d,j+1}^T, \ldots, \mathbf{h}_{d,J}^T, \mathbf{h}_{e,1}^T, \ldots, \mathbf{h}_{e,K}^T\right]^T$ and let $\mathbf{\Pi}_j^{\perp} \triangleq \mathbf{I} - \tilde{\mathbf{H}}_j (\tilde{\mathbf{H}}_j^H \tilde{\mathbf{H}}_j)^{-1} \tilde{\mathbf{H}}_j^H$ be the orthogonal complement projector of $\tilde{\mathbf{H}}_j$. In this case, the ZF beamformer can be expressed as $\mathbf{f}_j = \sqrt{P_j}\tilde{\mathbf{f}}_j$, where $\|\tilde{\mathbf{f}}_j\|^2 = 1$ and $\tilde{\mathbf{f}}_j = \mathbf{\Pi}_j^{\perp}\tilde{\mathbf{f}}_j$. By substituting the expression of $\mathbf{f}_j$ into (3.45a), the optimization problem can be reformulated as

$$\min_{\tilde{\mathbf{f}}_j, \forall j} \sum_{j=1}^{J} P_j \tag{3.48a}$$

$$\text{subject to } 1 + P_j |\mathbf{h}_{d,j}\mathbf{\Pi}_j^{\perp}\tilde{\mathbf{f}}_j|^2 \geq 2^{R_{0,j}}, \tag{3.48b}$$

$$\tilde{\mathbf{f}}_j = \mathbf{\Pi}_j^{\perp}\tilde{\mathbf{f}}_j, \ \|\tilde{\mathbf{f}}_j\|^2 = 1, \ \forall j. \tag{3.48c}$$

From the constraint in (3.48b), we can see that the power P_j is minimized when $\tilde{\mathbf{f}}_j$ is chosen to maximize the term $|\mathbf{h}_{d,j}\mathbf{\Pi}_j^{\perp}\tilde{\mathbf{f}}_j|^2$. By the Cauchy-Schwarz inequality, it follows that $|\mathbf{h}_{d,j}\mathbf{\Pi}_j^{\perp}\tilde{\mathbf{f}}_j|^2 \leq \|\mathbf{\Pi}_j^{\perp}\mathbf{h}_{d,j}^H\|^2\|\tilde{\mathbf{f}}_j\|^2$, which holds with equality when $\tilde{\mathbf{f}}_j = c\mathbf{\Pi}_j^{\perp}\mathbf{h}_{d,j}^H$, for some arbitrary constant c. Hence, the optimal ZF beamformer that minimizes the sum power is given by

$$\mathbf{f}_j^* = \frac{\sqrt{2^{R_{0,j}-1}}}{\|\mathbf{\Pi}_j^{\perp}\mathbf{h}_{d,j}^H\|^2}\mathbf{\Pi}_j^{\perp}\mathbf{h}_{d,j}^H \tag{3.49}$$

for $j = 1, \ldots, J$. It is worthwhile to remark that the ZF approach is generally suboptimal since it may restrict the source from transmitting in a direction that yields high channel gains to the intended destination.

Furthermore, it is worthwhile to remark that many variants of the multiuser secrecy beamforming or precoding problems can be considered. For example, one can consider the case with multiple sources transmitting to a single destination [25, 26] (i.e., the multiple access channel scenario), the case with multiple sources transmitting to multiple destinations [27] (i.e., the interference channel scenario), and the case of colluding eavesdroppers [28]. Moreover, in the above discussions, we focused only on the case where full and perfect channel state information (CSI) is available at the source. This is often not the case in practice. In fact, the CSI at the source is typically subject to errors caused by noisy channel estimation or limited feedback. In this case, robust beamforming and precoding schemes can be designed by taking into consideration deterministic or stochastic error bounds, e.g., in [12]. However, in these cases, pure beamforming may not be the best choice since an undesirable amount of information may leak into the eavesdropper's channel when perfect CSI is not available. The use of artificial noise may then be necessary to construct a favorable secrecy channel.

3.3 Secrecy Beamforming and Precoding with Artificial Noise

In addition to using beamforming and precoding schemes to strengthen (or weaken) signals in certain dimensions, it is often useful to emit artificial noise (AN) on top

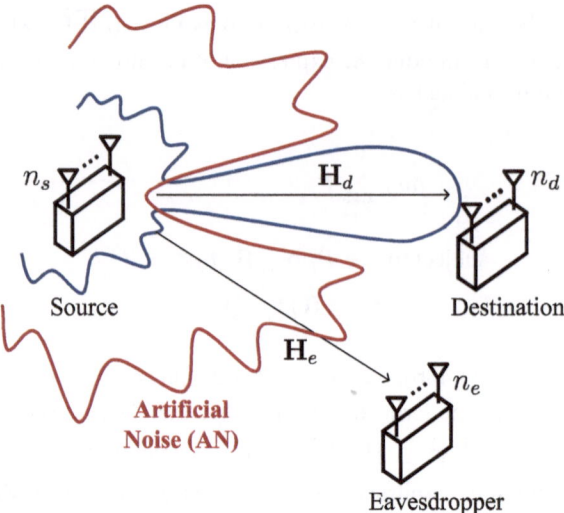

Fig. 3.3 An illustration of secrecy beamforming with artificial noise

of the information-bearing signal to disrupt the reception at the eavesdropper, as depicted in Fig. 3.3. This approach was first proposed in [29, 30] and is often used when the eavesdropper CSI is not perfectly known at the source since, in this case, the signals cannot be accurately placed in dimensions that are less favorable to the eavesdropper. Recall that, if perfect knowledge of both the destination and the eavesdropper CSI is available at the source, it is optimal to transmit signals in subchannels corresponding to the maximum generalized eigenvalue of the channel (cf. Sect. 3.1) and, thus, no AN is needed. However, in practice, the eavesdropper's CSI cannot be perfectly obtained due to channel estimation and feedback errors and, in fact, may not be available at all if the eavesdropper is a pure adversary and does not request any service from the source. The design of secrecy beamforming and precoding with AN are discussed in this section.

3.3.1 Secrecy Beamforming with Artificial Noise

Let us consider a basic multi-antenna wiretap channel that consists of a source with n_s antennas, a single-antenna destination, and an eavesdropper with n_e antennas. In the AN-assisted secrecy beamforming scheme, the signal transmitted by the source can be written in general as

$$\mathbf{x} = \mathbf{s} + \mathbf{a}, \qquad (3.50)$$

where $\mathbf{s} \in \mathbb{C}^{n_s \times 1}$ is the information-bearing signal and $\mathbf{a} \in \mathbb{C}^{n_s \times 1}$ is the AN vector with covariance matrix $\mathbf{K_a} \triangleq \mathrm{E}[\mathbf{aa}^H]$. Here, $\mathbf{a}$ is assumed to be independent of $\mathbf{s}$. The information bearing signal can be written as $\mathbf{s} = \mathbf{f}u$, where $\mathbf{f}$ is the beamforming vector and u is the encoded symbol. The signals received at the destination and the eavesdropper are given by

$$y = \mathbf{h}_d \mathbf{x} + w, \tag{3.51a}$$

$$\mathbf{z} = \mathbf{H}_e \mathbf{x} + \mathbf{v}. \tag{3.51b}$$

where $\mathbf{h}_d \in \mathbb{C}^{1 \times n_s}$ and $\mathbf{H}_e \in \mathbb{C}^{n_e \times n_s}$ are the channels to the destination and the eavesdropper, respectively, and $w \sim \mathcal{CN}(0, 1)$ and $\mathbf{v} \sim \mathcal{CN}(\mathbf{0}, \mathbf{I}_{n_e})$ are the AWGN. It is worthwhile to note that, from an information-theoretic perspective, the AN-assisted secrecy beamforming scheme can be viewed as a special case of the channel prefixing technique mentioned in Chap. 2. Here, the effective channel input is given by $u \sim p_u$ and the channel prefixing distribution is given by $p_{\mathbf{x}|u} \sim \mathcal{CN}(\mathbf{f}u, \mathbf{K}_a)$.

Let us consider the case where the eavesdropper CSI is not available at the source. This is often the case in practice since the eavesdropper is an adversary and usually would not be willing to reveal extra information to the source. In this case, the information-bearing signal can only be directed towards the destination to maximize its receive signal quality. This is achieved by choosing the beamforming vector to be

$$\mathbf{f} = \sqrt{P_s} \frac{\mathbf{h}_d^H}{\|\mathbf{h}_d\|}, \tag{3.52}$$

where P_s is the power of the signal component. Moreover, to prevent AN interference at the destination, AN should be placed in the null space of the destination's channel, i.e., $\mathbf{h}_d$, and can be expressed as

$$\mathbf{a} = \mathbf{N}_{\mathbf{h}_d} \tilde{\mathbf{a}}, \tag{3.53}$$

where $\mathbf{N}_{\mathbf{h}_d} \in \mathbb{C}^{n_s \times (n_s - 1)}$ is a matrix whose columns form an orthonormal basis for the null space of $\mathbf{h}_d$ such that $\mathbf{h}_d \mathbf{N}_{\mathbf{h}_d} = \mathbf{0}$ and $\mathbf{N}_{\mathbf{h}_d}^H \mathbf{N}_{\mathbf{h}_d} = \mathbf{I}_{n_s - 1}$, and $\tilde{\mathbf{a}} \in \mathbb{C}^{(n_s - 1) \times 1}$ is a Gaussian vector with distribution $\mathcal{CN}(\mathbf{0}, \sigma_a^2 \mathbf{I}_{n_s - 1})$. Notice that AN is scattered isotropically in the null space of the main channel since the eavesdropper channel is assumed to be completely unknown. The AN vector can be equivalently written as the projection of an isotropically distributed Gaussian random vector in $\mathbb{C}^{n_s}$ to the null space of $\mathbf{h}_d$, i.e.,

$$\mathbf{a} = \mathbf{\Pi}_{\mathbf{h}_d^H}^\perp \check{\mathbf{a}}, \tag{3.54}$$

where

$$\mathbf{\Pi}_{\mathbf{h}_d^H}^\perp \triangleq \mathbf{I}_{n_s} - \mathbf{h}_d^H (\mathbf{h}_d \mathbf{h}_d^H)^{-1} \mathbf{h}_d = \mathbf{I}_{n_s} - \frac{\mathbf{h}_d^H}{\|\mathbf{h}_d\|} \frac{\mathbf{h}_d}{\|\mathbf{h}_d\|}$$

is the orthogonal complement projector of $\mathbf{h}_d^H$ and $\check{\mathbf{a}} \sim \mathcal{CN}(\mathbf{0}, \sigma_a^2 \mathbf{I}_{n_s})$ is an $n_s \times 1$ Gaussian vector. Hence, the AN covariance matrix can be expressed as either $\mathbf{K}_{\mathbf{a}} = \sigma_a^2 \mathbf{N}_{\mathbf{h}_d} \mathbf{N}_{\mathbf{h}_d}^H$ or $\mathbf{K}_{\mathbf{a}} = \sigma_a^2 \mathbf{\Pi}_{\mathbf{h}_d^H}^\perp (\mathbf{\Pi}_{\mathbf{h}_d^H}^\perp)^H$. Here, AN is chosen to be Gaussian and is assumed to vary independently in each symbol period so that it is not easily predictable by the eavesdropper. By adopting the choice of $\mathbf{f}$ and $\mathbf{a}$ given in (3.52) and (3.53) (or (3.54)), the received signals at the destination and the eavesdropper can be written as

$$y = \mathbf{h}_d(fu + \mathbf{a}) + w = \sqrt{P_s}\|\mathbf{h}_d\|u + w \qquad (3.55)$$

$$\mathbf{z} = \mathbf{H}_e(fu + \mathbf{a}) + \mathbf{v} = \sqrt{P_s}\mathbf{H}_e\frac{\mathbf{h}_d^H}{\|\mathbf{h}_d\|}u + \underbrace{\mathbf{H}_e\mathbf{a} + \mathbf{v}}_{\mathbf{v}'}. \qquad (3.56)$$

Here, $\mathbf{v}' \triangleq \mathbf{H}_e\mathbf{a} + \mathbf{v}$ is defined as the equivalent noise at the eavesdropper and can be shown to be Gaussian with zero mean and covariance matrix $\mathbf{K}_{\mathbf{v}'} = E[\mathbf{v}'(\mathbf{v}')^H] = \mathbf{H}_e\mathbf{K}_{\mathbf{a}}\mathbf{H}_e^H + \mathbf{I}_{n_e}$. With $\mathbf{D}$ such that $\mathbf{K}_{\mathbf{v}'}^{-1} = \mathbf{D}\mathbf{D}^H$, the equivalent channel output at the eavesdropper can be written as

$$\tilde{\mathbf{z}} \triangleq \mathbf{D}^H\mathbf{z} = \sqrt{P_s}\mathbf{D}^H\mathbf{H}_e\frac{\mathbf{h}_d^H}{\|\mathbf{h}_d\|}u + \mathbf{D}^H\mathbf{v}' = \sqrt{P_s}\tilde{\mathbf{h}}_e u + \tilde{\mathbf{v}}, \qquad (3.57)$$

where $\tilde{\mathbf{h}}_e \triangleq \mathbf{D}^H\mathbf{H}_e\mathbf{h}_d^H/\|\mathbf{h}_d\|$ is the equivalent channel vector and $\tilde{\mathbf{v}} \triangleq \mathbf{D}^H\mathbf{v}' = \mathbf{D}^H\mathbf{H}_e\mathbf{a} + \mathbf{D}^H\mathbf{v}$ is the equivalently whitened noise, since $\tilde{\mathbf{v}}$ is now zero mean with covariance matrix $\mathbf{K}_{\tilde{\mathbf{v}}} = \mathbf{I}_{n_s}$. By taking $u \sim \mathcal{CN}(0, 1)$ as the equivalent channel input, it follows from Theorem 2.4 that the secrecy rate achievable with the AN-assisted beamforming scheme can be given as

$$R_s = \log\frac{1 + P_s\|\mathbf{h}_d\|^2}{\det(\mathbf{I}_{n_e} + P_s\tilde{\mathbf{h}}_e\tilde{\mathbf{h}}_e^H)} = \log\frac{1 + P_s\|\mathbf{h}_d\|^2}{\det(\mathbf{I}_{n_e} + P_s\mathbf{H}_e\frac{\mathbf{h}_d^H}{\|\mathbf{h}_d\|}\frac{\mathbf{h}_d}{\|\mathbf{h}_d\|}\mathbf{H}_e^H\mathbf{K}_{\mathbf{v}'}^{-1})}, \qquad (3.58)$$

where the second equality follows from Sylvester's determinant theorem [5], i.e., $\det(\mathbf{I}_m + \mathbf{AB}) = \det(\mathbf{I}_k + \mathbf{BA})$, for any $\mathbf{A} \in \mathbb{C}^{m\times k}$ and $\mathbf{B} \in \mathbb{C}^{k\times m}$. By adopting the AN model in (3.54), the term in the denominator inside the logarithm of (3.58) can be written as

$$\det\left(\mathbf{I}_{n_e} + P_s\mathbf{H}_e\frac{\mathbf{h}_d^H}{\|\mathbf{h}_d\|}\frac{\mathbf{h}_d}{\|\mathbf{h}_d\|}\mathbf{H}_e^H\mathbf{K}_{\mathbf{v}'}^{-1}\right)$$

$$= \det\left(\mathbf{I}_{n_e} + P_s\mathbf{H}_e\frac{\mathbf{h}_d^H}{\|\mathbf{h}_d\|}\frac{\mathbf{h}_d}{\|\mathbf{h}_d\|}\mathbf{H}_e^H\left[\mathbf{I}_{n_e} + \sigma_a^2\mathbf{H}_e\left(\mathbf{I}_{n_s} - \frac{\mathbf{h}_d^H}{\|\mathbf{h}_d\|}\frac{\mathbf{h}_d}{\|\mathbf{h}_d\|}\right)\mathbf{H}_e^H\right]^{-1}\right). \qquad (3.59)$$

This shows that, by increasing the AN variance σ_a^2, the denominator inside the logarithm of (3.58) decreases and, thus, the achievable secrecy rate is increased.

Let us consider the special case where $P_s = \sigma_a^2 = \bar{P}/n_s$. In this case, the total average transmit power is given by $E[\|\mathbf{x}\|^2] = E[\|\mathbf{s}\|^2] + E[\|\mathbf{a}\|^2] = P_s + (n_s -$

1)$\sigma_a^2 = \bar{P}$. Then, the term in (3.59) can be written as

$$
\det\left(\mathbf{I}_{n_e} + \frac{\bar{P}}{n_s}\mathbf{H}_e\frac{\mathbf{h}_d^H}{\|\mathbf{h}_d\|}\frac{\mathbf{h}_d}{\|\mathbf{h}_d\|}\mathbf{H}_e^H\left[\mathbf{I}_{n_e} + \frac{\bar{P}}{n_s}\mathbf{H}_e\left(\mathbf{I}_{n_s} - \frac{\mathbf{h}_d^H}{\|\mathbf{h}_d\|}\frac{\mathbf{h}_d}{\|\mathbf{h}_d\|}\right)\mathbf{H}_e^H\right]^{-1}\right)
$$

$$
= \det\left(\mathbf{I}_{n_e} + \frac{\bar{P}}{n_s}\mathbf{H}_e\mathbf{H}_e^H\right)\det\left(\mathbf{I}_{n_e} + \frac{\bar{P}}{n_s}\mathbf{H}_e\left(\mathbf{I}_{n_s} - \frac{\mathbf{h}_d^H}{\|\mathbf{h}_d\|}\frac{\mathbf{h}_d}{\|\mathbf{h}_d\|}\right)\mathbf{H}_e^H\right)^{-1}
$$

$$
= \det\left(\mathbf{I}_{n_s} + \frac{\bar{P}}{n_s}\mathbf{H}_e^H\mathbf{H}_e\right)\det\left(\mathbf{I}_{n_s} + \frac{\bar{P}}{n_s}\mathbf{H}_e^H\mathbf{H}_e - \frac{\bar{P}}{n_s}\frac{\mathbf{h}_d^H}{\|\mathbf{h}_d\|}\frac{\mathbf{h}_d}{\|\mathbf{h}_d\|}\mathbf{H}_e^H\mathbf{H}_e\right)^{-1}
$$

$$
= \det\left(\mathbf{I}_{n_s} - \frac{\bar{P}}{n_s}\frac{\mathbf{h}_d^H}{\|\mathbf{h}_d\|}\frac{\mathbf{h}_d}{\|\mathbf{h}_d\|}\mathbf{H}_e^H\mathbf{H}_e\left(\mathbf{I}_{n_s} + \frac{\bar{P}}{n_s}\mathbf{H}_e^H\mathbf{H}_e\right)^{-1}\right)^{-1}
$$

$$
= \left[\frac{\mathbf{h}_d}{\|\mathbf{h}_d\|}\left(\mathbf{I}_{n_s} + \frac{\bar{P}}{n_s}\mathbf{H}_e^H\mathbf{H}_e\right)^{-1}\frac{\mathbf{h}_d^H}{\|\mathbf{h}_d\|}\right]^{-1}. \tag{3.60}
$$

Hence, the achievable secrecy rate for this special case can be written as

$$
R_s = \log\left[\left(1 + \frac{\bar{P}}{n_s}\|\mathbf{h}_d\|^2\right)\frac{\mathbf{h}_d}{\|\mathbf{h}_d\|}\left(\mathbf{I}_{n_s} + \frac{\bar{P}}{n_s}\mathbf{H}_e^H\mathbf{H}_e\right)^{-1}\frac{\mathbf{h}_d^H}{\|\mathbf{h}_d\|}\right]
$$

$$
= \log\left(\frac{1}{\frac{\bar{P}}{n_s}\|\mathbf{h}_d\|^2} + 1\right) + \log\left[\frac{\bar{P}}{n_s}\mathbf{h}_d\left(\mathbf{I}_{n_s} + \frac{\bar{P}}{n_s}\mathbf{H}_e^H\mathbf{H}_e\right)^{-1}\mathbf{h}_d^H\right]
$$

$$
= \log\left(\frac{1}{\frac{\bar{P}}{n_s}\|\mathbf{h}_d\|^2} + 1\right) + \log\lambda_{\max}\left(\frac{\bar{P}}{n_s}\mathbf{h}_d^H\mathbf{h}_d, \mathbf{I}_{n_s} + \frac{\bar{P}}{n_s}\mathbf{H}_e^H\mathbf{H}_e\right), \tag{3.61}
$$

where $\lambda_{\max}(\mathbf{A}, \mathbf{B})$ denotes the maximum generalized eigenvalue of the matrix pair $(\mathbf{A}, \mathbf{B})$. Notice that, in the high SNR regime (i.e., when $\bar{P}/n_s$ is large), the achievable secrecy rate can be approximated as

$$
R_{s,\text{AN-BF}}(\bar{P}) \approx \log\lambda_{\max}\left(\mathbf{I}_{n_s} + \frac{\bar{P}}{n_s}\mathbf{h}_d^H\mathbf{h}_d, \mathbf{I}_{n_s} + \frac{\bar{P}}{n_s}\mathbf{H}_e^H\mathbf{H}_e\right). \tag{3.62}
$$

Recall from (3.13) that the right-hand-side of (3.62) is equal to the secrecy capacity, $C_s(\bar{P}/n_s)$, under the total power constraint $\bar{P}/n_s$ for the case where perfect knowledge of both the destination and the eavesdropper channels is available at the source. This is to say that a loss of $10\log_{10} n_s$ is experienced by adopting the AN-assisted beamforming scheme. This loss is expected since the design of the AN-assisted secrecy beamforming scheme does not utilize knowledge of the eavesdropper's channel.

It is necessary to remark that the secrecy rate in (3.58) is achieved by assuming that the coherence interval is sufficiently long and by transmitting each codeword over only a single channel state. However, when the eavesdropper CSI is not available at the source, the achievable secrecy rate in each coherence interval cannot be derived instantaneously and, thus, the code-rate cannot be accurately selected to guarantee successful decoding at the destination and to ensure perfect secrecy against the eavesdropper. In this case, one can utilize the secrecy outage probability as the performance measure. In delay-tolerant applications, one can also encode over multiple channel states, in which case, the achievable ergodic secrecy rate is given by [31]

$$R_{s,\text{ergodic}} = \left\{ \mathbb{E}_{\mathbf{h}_d, \mathbf{H}_e} \left[\log \frac{1 + P_s \|\mathbf{h}_d\|^2}{1 + P_s \frac{\mathbf{h}_d}{\|\mathbf{h}_d\|} \mathbf{H}_e^H \mathbf{K}_{\mathbf{v}'}^{-1} \mathbf{H}_e \frac{\mathbf{h}_d^H}{\|\mathbf{h}_d\|}} \right] \right\}^+ . \tag{3.63}$$

where $\mathbf{K}_{\mathbf{v}'} = \mathbf{I}_{n_e} + \sigma_a^2 \mathbf{H}_e \left(\mathbf{I}_{n_s} - \frac{\mathbf{h}_d^H}{\|\mathbf{h}_d\|} \frac{\mathbf{h}_d}{\|\mathbf{h}_d\|} \right) \mathbf{H}_e^H .$

3.3.2 Power Allocation Between Information Signal and Artificial Noise

From the ergodic secrecy rate expression in (3.63), one can observe that, even though the use of AN can effectively deteriorate the eavesdropper channel, it also reduces the power that can be utilized for the transmission of the information signal. In this sense, the use of AN may indirectly reduce the SINR at the destination compared to the case with no AN. Therefore, power allocation between signal and AN is important to ensure good performance under secrecy constraints. These issues have been studied in [31] and are summarized below.

Specifically, let us consider the power allocation problem where the signal power, i.e., P_s, and the AN variance, i.e., σ_a^2, are chosen to maximize the ergodic secrecy rate in (3.63). The expectation in (3.63) is taken over $\mathbf{h}_d$ and $\mathbf{H}_e$ that are assumed to have i.i.d. entries with distributions $\mathcal{CN}(0, \sigma_{h_d}^2)$ and $\mathcal{CN}(0, \sigma_{h_e}^2)$, respectively. Similar to the previous section, we assume that the source has instantaneous knowledge of the realizations of $\mathbf{h}_d$, but has only the statistics of $\mathbf{H}_e$. The signal power and the AN variance are assumed to satisfy the total power constraint

$$P_s + (n_s - 1)\sigma_a^2 = \bar{P}, \tag{3.64}$$

and, thus, can be expressed as $P_s = \alpha \bar{P}$ and $\sigma_a^2 = \frac{(1-\alpha)\bar{P}}{n_s - 1}$, where α is the fraction of the total power allocated to the information signal. Here, we assume that power allocation is non-adaptive and does not vary over different channel states. If instantaneous knowledge of the destination's channel is available at the source, power allocation can also be adapted to the realization of $\mathbf{h}_d$. However, this is not discussed in this chapter. Readers are referred to [31] for further discussions on the design of adaptive power control strategies.

To derive the power allocation, let us consider the secrecy capacity lower bound which is obtained by assuming that the reception at the eavesdropper is noiseless. In this case, the received signal at the eavesdropper can be written as

$$\mathbf{z} = \mathbf{H}_e(\mathbf{f}u + \mathbf{a}) = \sqrt{P_s}\mathbf{H}_e\frac{\mathbf{h}_d^H}{\|\mathbf{h}_d\|}u + \mathbf{v}''. \tag{3.65}$$

The equivalent noise is now given by $\mathbf{v}'' \triangleq \mathbf{H}_e\mathbf{a}$, where the covariance matrix is given by $\mathbf{K}_{\mathbf{v}''} = \sigma_a^2\mathbf{H}_e\left(\mathbf{I}_{n_s} - \frac{\mathbf{h}_d^H}{\|\mathbf{h}_d\|}\frac{\mathbf{h}_d}{\|\mathbf{h}_d\|}\right)\mathbf{H}_e^H = \sigma_a^2\mathbf{H}_e\mathbf{N}_{\mathbf{h}_d}\mathbf{N}_{\mathbf{h}_d}^H\mathbf{H}_e^H$. For a given power allocation α, the ergodic secrecy rate in (3.63) can then be lower-bounded as

$$R_{s,\text{lower}}(\alpha) = \left\{\mathrm{E}_{\mathbf{h}_d}\left[\log(1 + \alpha\bar{P}\|\mathbf{h}_d\|^2)\right]\right.$$
$$\left. - \mathrm{E}_{\mathbf{h}_d,\mathbf{H}_e}\left[\log\left(1 + \frac{\alpha(n_s - 1)}{1 - \alpha}\mathbf{g}_1^H(\mathbf{G}_2\mathbf{G}_2^H)^{-1}\mathbf{g}_1\right)\right]\right\}^+. \tag{3.66}$$

where $\mathbf{g}_1 \triangleq \mathbf{H}_e\mathbf{h}_d^H/\|\mathbf{h}_d\|$ and $\mathbf{G}_2 \triangleq \mathbf{H}_e\mathbf{N}_{\mathbf{h}_d}$. Since $\mathbf{h}_d^H/\|\mathbf{h}_d\|$ and the columns of $\mathbf{N}_{\mathbf{h}_d}$ are orthonormal, the entries of $\mathbf{g}_1$ and $\mathbf{G}_2$ are also i.i.d. $\mathcal{CN}(0, \sigma_{h_e}^2)$. Notice that the secrecy capacity lower bound (3.66) can be viewed as the worst case performance and is a valid criterion when the noise-level at the eavesdropper is also unknown.

First of all, since the entries of $\mathbf{h}_d$ are i.i.d. $\mathcal{CN}(0, \sigma_{h_d}^2)$, it follows that $\|\mathbf{h}_d\|^2/\sigma_{h_d}^2$ follows a Gamma distribution with parameters $(n_s, 1)$. Therefore, the first term in (3.66) can be written in the following integral form [31]:

$$\mathrm{E}_{\mathbf{h}_d}\left[\log(1 + \alpha\bar{P}\|\mathbf{h}_d\|^2)\right] = \frac{1}{\ln 2}\int_0^\infty \ln(1 + \alpha\bar{P}\sigma_{h_d}^2 x)x^{n_s - 1}\frac{\exp(-x)}{\Gamma(n_s)}dx$$
$$= \frac{1}{\ln 2}\exp\left(\frac{1}{\alpha\bar{P}\sigma_{h_d}^2}\right)\sum_{k=1}^{n_s}E_k\left(\frac{1}{\alpha\bar{P}\sigma_{h_d}^2}\right), \tag{3.67}$$

where $\Gamma(\cdot)$ is the Gamma function and $E_k(\cdot)$ is the generalized exponential integral.

Secondly, by the fact that $\mathbf{g}_1$ and $\mathbf{G}_2$ have entries that are i.i.d. $\mathcal{CN}(0, \sigma_{h_e}^2)$, it was observed in [31] that the quantity $\mathbf{g}_1^H(\mathbf{G}_2\mathbf{G}_2^H)^{-1}\mathbf{g}_1$ can be viewed as the signal-to-interference ratio of an n_e-branch MMSE diversity combiner with $n_s - 1$ interferers. The complementary cumulative distribution function of $X = \mathbf{g}_1^H(\mathbf{G}_2\mathbf{G}_2^H)^{-1}\mathbf{g}_1$ is given in [32] as

$$Q_X(x) - \frac{1}{(1 + x)^{n_s - 1}}\sum_{k=0}^{n_e - 1}\binom{n_s - 1}{k}x^k. \tag{3.68}$$

The second term in (3.66) can then be written as [31]

$$
E_X\left[\log\left(1+\frac{\alpha(n_s-1)}{1-\alpha}X\right)\right]
$$

$$
=\frac{1}{\ln 2}\int_0^\infty \frac{\alpha(n_s-1)}{1-\alpha}\left(1+\frac{\alpha(n_s-1)}{1-\alpha}x\right)^{-1}Q_X(x)dx
$$

$$
=\frac{1}{\ln 2}\sum_{k=0}^{n_e-1}\binom{n_s-1}{k}\frac{\alpha(n_s-1)}{1-\alpha}B(k+1,n_s-1-k)\cdot {}_2F_1\left(1,k+1;n_s;\frac{1-\alpha n_s}{1-\alpha}\right)
$$

$$
\tag{3.69}
$$

where $B(a,b)=\Gamma(a)\Gamma(b)/\Gamma(a+b)$ is the beta function, and ${}_2F_1(\cdot)$ is the Gauss hypergeometric function. The first equality is obtained by using integration by parts whereas the second equality follows from an integration density in [33]. By substituting (3.67) and (3.69) into (3.66), the optimal power allocation can then be found numerically by performing a simple line search over $\alpha \in [0,1]$.

Interestingly, when the number of source antennas, i.e., n_s, is large, it follows from the law of large numbers that $\lim_{n_s\to\infty}\|\mathbf{h}_d\|^2/n_s = \sigma_{h_d}^2$ and $\lim_{n_s\to\infty}\mathbf{G}_2\mathbf{G}_2^H/(n_s-1)=\sigma_{h_e}^2\mathbf{I}_{n_e}$. Therefore, as $n_s \to \infty$, the first term in (3.66) can be approximated as

$$
E_{\mathbf{h}_d}\left[\log(1+\alpha\bar{P}\|\mathbf{h}_d\|^2)\right]=E_{\mathbf{h}_d}\left[\log\left[n_s\left(\frac{1}{n_s}+\alpha\bar{P}\frac{\|\mathbf{h}_d\|^2}{n_s}\right)\right]\right]
\tag{3.70}
$$

$$
\approx \log(\alpha\bar{P}\sigma_{h_d}^2 n_s)
\tag{3.71}
$$

and the second term can be approximated as

$$
E_{\mathbf{h}_d,\mathbf{H}_e}\left[\log\left(1+\frac{\alpha(n_s-1)}{1-\alpha}\mathbf{g}_1^H(\mathbf{G}_2\mathbf{G}_2^H)^{-1}\mathbf{g}_1\right)\right]
$$

$$
=E_{\mathbf{h}_d,\mathbf{H}_e}\left[\log\left(1+\frac{\alpha}{1-\alpha}\mathbf{g}_1^H\left(\frac{\mathbf{G}_2\mathbf{G}_2^H}{n_s-1}\right)^{-1}\mathbf{g}_1\right)\right]
$$

$$
\approx E_{\mathbf{h}_d,\mathbf{H}_e}\left[\log\left(1+\frac{\alpha}{1-\alpha}\frac{\|\mathbf{g}_1\|^2}{\sigma_{h_e}^2}\right)\right]
$$

$$
=\frac{1}{\ln 2}\exp\left(\frac{1-\alpha}{\alpha}\right)\sum_{k=1}^{n_s}E_k\left(\frac{1-\alpha}{\alpha}\right).
\tag{3.72}
$$

Hence, the secrecy capacity lower bound is approximated as

$$
R_{s,\text{lower}}(\alpha)\approx \log(\alpha\bar{P}\sigma_{h_d}^2 n_s)-\frac{1}{\ln 2}\exp\left(\frac{1-\alpha}{\alpha}\right)\sum_{k=1}^{n_s}E_k\left(\frac{1-\alpha}{\alpha}\right)
\tag{3.73}
$$

$$= \log(\alpha \bar{P} \sigma_{h_d}^2 n_s) - \frac{1}{\ln 2} \exp(z-1) \sum_{k=1}^{n_s} E_k(z-1) \tag{3.74}$$

$$\triangleq R_{s,\text{approx}}(z), \tag{3.75}$$

where $z \triangleq 1/\alpha$. By taking the derivative of $R_{s,\text{approx}}(z)$ with respect to z, the optimal z should satisfy

$$\frac{\partial R_{s,\text{approx}}(z)}{\partial z} = -\frac{1}{z} - \exp(z-1)E_{n_e}(z-1) + \frac{1}{z-1} = 0. \tag{3.76}$$

By using $\exp(z-1)E_{n_e}(z-1) \approx (z-1+n_e)^{-1}$ from [34], which is accurate when n_e or z is large, the optimal z is given by

$$z^* = 1 + \sqrt{n_e}. \tag{3.77}$$

The optimal power allocation α is thus given by [31]

$$\alpha^* = \frac{1}{1 + \sqrt{n_e}}. \tag{3.78}$$

This shows that, as n_e increases, the power allocated to AN should increase in order to provide the information signal with more protection.

It is worthwhile to remark that the study of signal and AN power allocation can also be extended to the case with imperfect CSI. In particular, when the CSI at the source is imperfect due to quantized feedback, it was shown in [35] that AN power should be allocated more conservatively since AN leakage into the main channel may cause significant performance degradation at the destination. Readers are referred to [35] for further details on these issues.

Example 3.1 In Fig. 3.4, the achievable ergodic secrecy rates are shown for systems with $n_s = 6, n_d = 1, n_e = 1, 5, 9$, and SNR $= 10$ dB. The solid curves represent the achievable ergodic secrecy rates obtained under the large n_s approximation (i.e., the secrecy rate given in (3.74)) with the asymptotically optimal power allocation policy given in (3.78). The dashed curves represent the exact achievable ergodic secrecy rate computed by substituting (3.67) and (3.69) into (3.66) with the asymptotically optimal power allocation given in (3.78). The dotted curve represents the case where the optimal power allocation is obtained by carrying out a line search over $\alpha \in (0, 1)$ based on the exact secrecy rate expression.

It can be observed that the exact secrecy rate expression indeed converges towards its asymptotic expression as n_s becomes large. Even though the asymptotic secrecy rate expression may not yield a close approximation for cases with small n_s, the asymptotic power allocation is sufficient to achieve good performance compared to that obtained via line search.

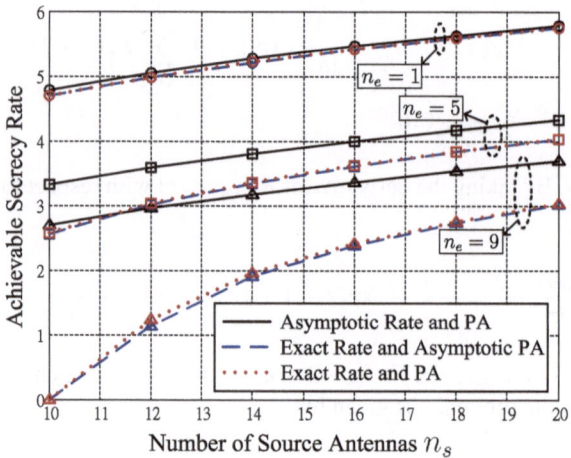

Fig. 3.4 Comparison of the achievable ergodic secrecy rates for MISOME channels with $n_s = 6$, $n_d = 1$, $n_e = 1, 5, 9$, and SNR $= 10$ dB

3.3.3 Secrecy Precoding with Artificial Noise

The use of AN to mask the transmission of information signals can also be extended to cases with multiple antennas at the destination, i.e., with $n_d > 1$. In this case, a higher achievable secrecy rate can be achieved by encoding over multiple spatial dimensions and, thus, AN should be placed in the null space of the dimensions corresponding to the information signal.

Specifically, let us consider a multi-antenna wiretap channel with n_s, n_d, and n_e antennas at the source, the destination, and the eavesdropper, respectively. For convenience, let us assume that $n_d \leq n_s \leq n_e$. The signal transmitted by the source can be written similarly as

$$\mathbf{x} = \mathbf{s} + \mathbf{a}. \tag{3.79}$$

The information signal $\mathbf{s}$ can now be written as

$$\mathbf{s} = \mathbf{Fu}, \tag{3.80}$$

where $\mathbf{F} \in \mathbb{C}^{n_s \times k_s}$ is the precoding matrix and $\mathbf{u} \in \mathbb{C}^{k_s \times 1}$ is the encoded symbol vector with $\mathrm{E}[\mathbf{uu}^H] = \mathbf{I}_{k_s}$. Similarly, to prevent AN interference at the destination, AN is placed in the null space of the destination's channel, $\mathbf{H}_d$, and can be expressed as

$$\mathbf{a} = \mathbf{N}_{\mathbf{H}_d} \tilde{\mathbf{a}} \tag{3.81}$$

where $\mathbf{N}_{\mathbf{H}_d}$ is a matrix whose columns form an orthonormal basis for the null space of $\mathbf{H}_d$ (i.e., $\mathbf{H}_d \mathbf{N}_{\mathbf{H}_d} = \mathbf{0}$ and $\mathbf{N}_{\mathbf{H}_d}^H \mathbf{N}_{\mathbf{H}_d} = \mathbf{I}_{n_s - n_d}$) and $\mathbf{a} \sim \mathcal{CN}(0, \sigma_a^2 \mathbf{I}_{n_s - n_d})$ is a Gaussian vector with i.i.d. entries. The AN vector can also be written equivalently as the projection of an i.i.d. Gaussian vector onto the null space of $\mathbf{H}_d$, i.e.,

$$\mathbf{a} = \mathbf{\Pi}^{\perp}_{\mathbf{H}_d^H} \check{\mathbf{a}}, \tag{3.82}$$

where $\mathbf{\Pi}^{\perp}_{\mathbf{H}_d^H} \triangleq \mathbf{I}_{n_s} - \mathbf{H}_d^H(\mathbf{H}_d\mathbf{H}_d^H)^{-1}\mathbf{H}_d$ is the orthogonal complement projector of $\mathbf{H}_d^H$ and $\check{\mathbf{a}} \sim \mathcal{CN}(\mathbf{0}, \sigma_a^2\mathbf{I}_{n_s})$ is an $n_s \times 1$ Gaussian vector. Therefore, the AN covariance matrix can be expressed as $\mathbf{K}_{\mathbf{a}} = \sigma_a^2 \mathbf{N}_{\mathbf{h}_d}\mathbf{N}_{\mathbf{h}_d}^H$ or, equivalently, as $\mathbf{K}_{\mathbf{a}} = \sigma_a^2 \mathbf{\Pi}^{\perp}_{\mathbf{h}_d^H}(\mathbf{\Pi}^{\perp}_{\mathbf{h}_d^H})^H$. The received signals at the destination and the eavesdropper can then be written as

$$\mathbf{y} = \mathbf{H}_d(\mathbf{Fu} + \mathbf{a}) + \mathbf{w} = \mathbf{H}_d\mathbf{Fu} + \mathbf{w} \tag{3.83}$$

$$\mathbf{z} = \mathbf{H}_e(\mathbf{Fu} + \mathbf{a}) + \mathbf{v} = \mathbf{H}_e\mathbf{Fu} + \mathbf{v}' \tag{3.84}$$

where $\mathbf{v}' \triangleq \mathbf{H}_e\mathbf{a} + \mathbf{v}$ is the equivalent noise that is Gaussian with zero mean and covariance matrix $\mathbf{K}_{\mathbf{v}'} = \mathbb{E}[\mathbf{v}'(\mathbf{v}')^H] = \mathbf{H}_e\mathbf{K}_{\mathbf{a}}\mathbf{H}_e^H + \mathbf{I}_{n_e}$. Following similar derivations as in (3.58), an achievable secrecy rate is given by

$$R_s = \log\frac{\det(\mathbf{I}_{n_d} + \mathbf{H}_d\mathbf{FF}^H\mathbf{H}_d^H)}{\det(\mathbf{I}_{n_e} + \mathbf{H}_e\mathbf{FF}^H\mathbf{H}_e^H\mathbf{K}_{\mathbf{v}'}^{-1})}. \tag{3.85}$$

More specifically, let us take the compact singular value decomposition (SVD) of $\mathbf{H}_d$ as

$$\mathbf{H}_d = \mathbf{U}_d\mathbf{\Delta}_d\mathbf{V}_d^H, \tag{3.86}$$

where $\mathbf{U}_d \in \mathbb{C}^{n_d \times n_d}$ is a unitary matrix, $\mathbf{\Delta}_d \in \mathbb{C}^{n_d \times n_d}$ is a diagonal matrix with positive diagonal elements, and $\mathbf{V}_d \in \mathbb{C}^{n_s \times n_d}$ is a matrix with orthonormal columns. Let $\mathbf{v}_1, \ldots, \mathbf{v}_{n_d}$ be the columns of $\mathbf{V}_d$, i.e., $\mathbf{V}_d = [\mathbf{v}_1, \ldots, \mathbf{v}_{n_d}]$, and let $\mathbf{N}_{\mathbf{V}_d} = [\mathbf{v}_{n_d+1}, \ldots, \mathbf{v}_{n_s}]$ be a matrix whose columns form an orthonormal basis for the null space of $\mathbf{V}_d$. Following [3], let us consider the case where $k_s = n_d$ and let

$$\mathbf{F} = \sqrt{P_s}\mathbf{V}_d \tag{3.87}$$

and $\mathbf{N}_{\mathbf{H}_d} = \mathbf{N}_{\mathbf{V}_d}$. Then, by choosing $P_s = \sigma_a^2 = \bar{P}/n_s$ and by following derivations similar to that in (3.60) and (3.61), the achievable secrecy rate in (3.85) can be computed as

$$R_s = \log\det\left(\mathbf{I}_{n_d} + \frac{\bar{P}}{n_s}\mathbf{\Delta}_d^2\right) + \log\det\left(\mathbf{V}_d^H\left(\mathbf{I}_{n_s} + \frac{\bar{P}}{n_s}\mathbf{H}_e^H\mathbf{H}_e\right)^{-1}\mathbf{V}_d\right)$$

$$= \log\det\left[\left(\mathbf{\Delta}_d^{-2} + \frac{\bar{P}}{n_s}\mathbf{I}_{n_d}\right)\left(\mathbf{U}_d\mathbf{\Delta}_d\mathbf{V}_d^H\left(\mathbf{I}_{n_s} + \frac{\bar{P}}{n_s}\mathbf{H}_e^H\mathbf{H}_e\right)^{-1}\mathbf{V}_d\mathbf{\Delta}_d\mathbf{U}_d^H\right)\right]$$

$$= \log \det \left[\left(\Delta_d^{-2} + \frac{\bar{P}}{n_s} \mathbf{I}_{n_d} \right) \left(\mathbf{H}_d \left(\mathbf{I}_{n_s} + \frac{\bar{P}}{n_s} \mathbf{H}_e^H \mathbf{H}_e \right)^{-1} \mathbf{H}_d^H \right) \right] \qquad (3.88)$$

Let us take the generalized SVD of $\mathbf{H}_d$ and $\mathbf{H}_e$ [7, 8] such that

$$\mathbf{H}_d = \mathbf{\Psi}_d \mathbf{\Sigma}_d [\mathbf{\Omega}^{-1} \ \mathbf{0}_{k \times (n_s - k)}] \mathbf{\Psi}_s^H \qquad (3.89)$$

$$\mathbf{H}_e = \mathbf{\Psi}_e \mathbf{\Sigma}_e [\mathbf{\Omega}^{-1} \ \mathbf{0}_{k \times (n_s - k)}] \mathbf{\Psi}_s^H, \qquad (3.90)$$

where $\mathbf{\Psi}_d \in \mathbb{C}^{n_d \times n_d}$, $\mathbf{\Psi}_e \in \mathbb{C}^{n_e \times n_e}$, and $\mathbf{\Psi}_s \in \mathbb{C}^{n_s \times n_s}$ are unitary matrices, $\mathbf{\Omega} \in \mathbb{C}^{k \times k}$ is a lower triangular nonsingular matrix. Also, for $k = \mathrm{rank}([\mathbf{H}_d^H \ \mathbf{H}_e^H])$, $r = \dim(\mathrm{Null}(\mathbf{H}_d)^\perp \cap \mathrm{Null}(\mathbf{H}_e))$, and $s = \dim(\mathrm{Null}(\mathbf{H}_d)^\perp \cap \mathrm{Null}(\mathbf{H}_e)^\perp)$,

$$\mathbf{\Sigma}_d = \begin{pmatrix} \mathbf{0} & \mathbf{0} & \mathbf{0} \\ \mathbf{0} & \mathbf{D}_d & \mathbf{0} \\ \mathbf{0} & \mathbf{0} & \mathbf{I}_r \end{pmatrix} \qquad (3.91)$$

and

$$\mathbf{\Sigma}_e = \begin{pmatrix} \mathbf{I}_{k-r-s} & \mathbf{0} & \mathbf{0} \\ \mathbf{0} & \mathbf{D}_e & \mathbf{0} \\ \mathbf{0} & \mathbf{0} & \mathbf{0} \end{pmatrix} \qquad (3.92)$$

are $n_d \times k$ and $n_e \times k$ matrices, where $\mathbf{D}_d = \mathrm{diag}(\sigma_{d,1}, \ldots, \sigma_{d,s})$ and $\mathbf{D}_e = \mathrm{diag}(\sigma_{e,1}, \ldots, \sigma_{e,s})$ are diagonal matrices whose diagonal entries are real and strictly positive. The associated generalized singular values are

$$\sigma_i = \frac{\sigma_{d,i}}{\sigma_{e,i}}, \qquad (3.93)$$

for $i = 1, \ldots, n_d$, where the indices are chosen such that $\sigma_1 \le \sigma_2 \le \cdots \le \sigma_s$. In the high-SNR regime, the achievable secrecy rate reduces to [3]

$$\lim_{\bar{P} \to \infty} R_s(\bar{P}) = \log \det(\mathbf{H}_d (\mathbf{H}_e^H \mathbf{H}_e)^{-1} \mathbf{H}_d^H) = \sum_{i=1}^{n_d} \log \sigma_i^2. \qquad (3.94)$$

Example 3.2 The achievable secrecy rate is shown for three schemes in Fig. 3.5. They are: (i) secrecy beamforming (or precoding) without AN under perfect CSI, (ii) secrecy beamforming (or precoding) without AN under imperfect CSI, and (iii) secrecy beamforming (or precoding) with AN under imperfect CSI. Both the MIS-OME case (with $n_s = 6$, $n_d = 1$, $n_e = 2$) and the MIMOME case (with $n_s = 6$, $n_d = 2$, $n_e = 2$) are shown. The total power constraint P is adopted for the MISOME case while the power covariance constraint with $\mathbf{S} = \frac{P}{n_s} \mathbf{I}_{n_s}$ is used for the MIMOME case. Note that such power covariance constraint serves as a special case of the total power constraint P. For the case with imperfect CSI, it is assumed that only outdated

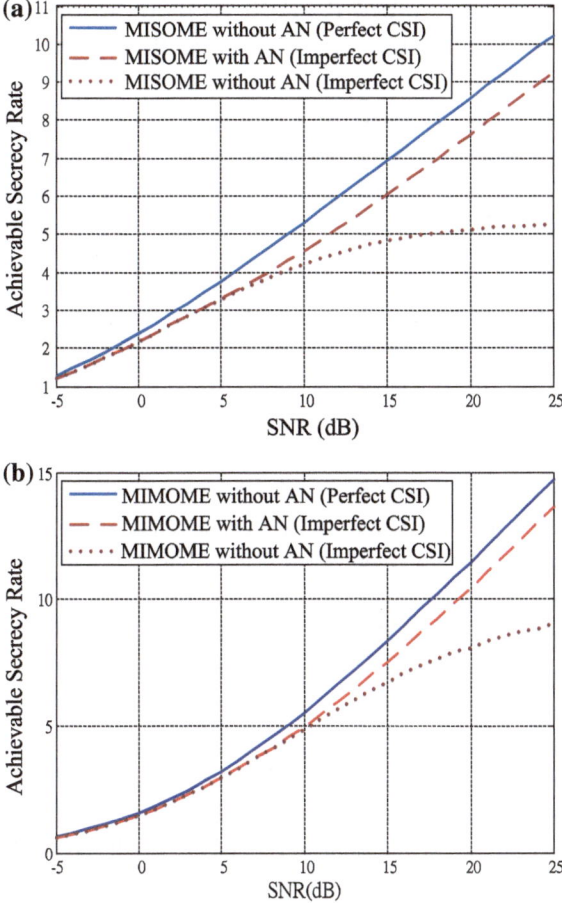

Fig. 3.5 Comparison of the achievable secrecy rate for (**a**) MISOME (i.e., $n_s = 6, n_d = 1, n_e = 2$) and (**b**) MIMOME (i.e., $n_s = 6, n_d = 2, n_e = 2$) cases with and without AN under different levels of CSI

channel information is available and in the form of

$$H'_d = (1 - \alpha)H_d + \alpha \Delta H_d;$$
$$H'_e = (1 - \alpha)H_e + \alpha \Delta H_e,$$

where α is set to 0.2 and the entries of H_d, ΔH_d, H_e, and ΔH_e are assumed to be i.i.d. $\mathcal{CN}(0, 1)$. We assume flat fading where the channel stays the same for a whole transmission block. All the AN is isotropic. One can observe that, in both MISOME and MIMOME scenarios, AN is essential to achieve high secrecy rate under imperfect CSI. However, when perfect CSI is available, pure secrecy beamforming or precoding is sufficient and the use of AN would be unnecessary.

3.4 Secrecy Beamforming with Artificial Noise for Systems with Multiple Destinations and Eavesdroppers

The use of AN to protect or mask the transmission of information-bearing signals can also be extended to systems with multiple destinations and eavesdroppers. Let us consider, as an example, the broadcast scenario where multiple data-streams are to be transmitted simultaneously to different destinations [15]. Similar to Sect. 3.2.2, we consider here a system that consists of a source with n_s antennas, J single-antenna destinations, and K single antenna eavesdroppers. Assume that $n_s > J$. Let u_j be the encoded symbol intended for destination j, where $E[|u_j|^2] = 1$, and let $\mathbf{f}_j$ be the corresponding beamforming vector. The signal transmitted by the source can then be written as

$$\mathbf{x} = \sum_{j=1}^{J} \mathbf{f}_j u_j + \mathbf{a}, \tag{3.95}$$

where $\mathbf{a} \in \mathbb{C}^{n_s \times 1}$ is the AN vector with zero mean and covariance matrix $\mathbf{K_a}$. The signal received at destination j and eavesdropper k are given by

$$y_j = \mathbf{h}_{d,j}\mathbf{f}_j u_j + \mathbf{h}_{d,j} \sum_{\ell \neq j} \mathbf{f}_\ell u_\ell + \mathbf{h}_{d,j}\mathbf{a} + w_j \tag{3.96}$$

and

$$z_k = \mathbf{h}_{e,k}\mathbf{f}_j u_j + \mathbf{h}_{e,k} \sum_{\ell \neq j} \mathbf{f}_\ell u_\ell + \mathbf{h}_{e,k}\mathbf{a} + v_k, \tag{3.97}$$

respectively, where $\mathbf{h}_{d,j} \in \mathbb{C}^{1 \times n_s}$ and $\mathbf{h}_{e,k} \in \mathbb{C}^{1 \times n_s}$ are the channels to destination j and eavesdropper k, respectively, and $w_j, v_k \sim \mathcal{CN}(0, 1)$ are the AWGN. Similar to (3.44), an achievable secrecy rate at destination j is given by

$$R_{s,j} = \log \frac{1 + \dfrac{\mathbf{h}_{d,j}\mathbf{K}_j\mathbf{h}_{d,j}^H}{1+\mathbf{h}_{d,j}\mathbf{K_a}\mathbf{h}_{d,j}^H+\sum_{\ell \neq j} \mathbf{h}_{d,j}\mathbf{K}_\ell\mathbf{h}_{d,j}^H}}{1 + \max_k \dfrac{\mathbf{h}_{e,k}\mathbf{K}_j\mathbf{h}_{e,k}^H}{1+\mathbf{h}_{e,k}\mathbf{K_a}\mathbf{h}_{e,k}^H+\sum_{\ell \neq j} \mathbf{h}_{e,k}\mathbf{K}_\ell\mathbf{h}_{e,k}^H}} \tag{3.98}$$

where $\mathbf{K}_j = \mathbf{f}_j\mathbf{f}_j^H$ is the rank-1 covariance matrix corresponding to the signal intended for destination j, i.e., $\mathbf{f}_j u_j$.

In general, the optimal beamforming vectors $\{\mathbf{f}_j\}_{j=1}^{J}$ and AN covariance matrix $\mathbf{K_a}$ can be found using the approach given in [15] and in Sect. 3.2.2. However, let us first consider, as an intuitive example, the ZF approach where both inter-user interference and AN are both nulled out at each destination. In particular, let us define $\mathbf{H} \triangleq [\mathbf{h}_{d,1}^H, \ldots, \mathbf{h}_{d,J}^H]^H \in \mathbb{C}^{J \times n_s}$ as the collection of channel vectors to the J destinations and assume that $\mathbf{H}$ is full rank. By taking the ZF approach, we can choose $\mathbf{f}_j$ such that $\mathbf{h}_{d,i}\mathbf{f}_j = 0$, for all $i \neq j$, and AN vector $\mathbf{a} = \mathbf{N_H}\tilde{\mathbf{a}}$, where

$\tilde{\mathbf{a}} \sim \mathcal{CN}(\mathbf{0}, \sigma_a^2 \mathbf{I}_{n_s-J})$ and $\mathbf{N_H} \in \mathbb{C}^{n_s \times (n_s-J)}$ is a matrix whose columns form an orthonormal basis for the null space of $\mathbf{H}$. Assume that equal power is allocated for the transmission of each destination's message, i.e., $\|\mathbf{f}_j\|^2 = P_s$, for all j. By letting $P_s = \alpha \bar{P}/J$ and $\sigma_a^2 = (1 - \alpha)\bar{P}/(n_s - J)$ such that the total transmit power is $JP_s + (n_s - J)\sigma_a^2 = \bar{P}$, the achievable secrecy rate in (3.98) reduces to

$$
R_{s,j} = \log \frac{1 + \mathbf{h}_{d,j}\mathbf{K}_j\mathbf{h}_{d,j}^H}{1 + \max\limits_k \dfrac{\mathbf{h}_{e,k}\mathbf{K}_j\mathbf{h}_{e,k}^H}{1+\mathbf{h}_{e,k}\mathbf{K_a}\mathbf{h}_{e,k}^H+\sum_{\ell \neq j}\mathbf{h}_{e,k}\mathbf{K}_\ell\mathbf{h}_{e,k}^H}}
$$

$$
\geq \log\left(1 + \frac{\alpha\bar{P}}{J}\mathbf{h}_{d,j}\tilde{\mathbf{f}}_j\tilde{\mathbf{f}}_j^H\mathbf{h}_{d,j}^H\right) - \log\left(\max\limits_k \frac{\alpha(n_s - J)\mathbf{h}_{e,k}\tilde{\mathbf{f}}_j\tilde{\mathbf{f}}_j^H\mathbf{h}_{e,k}^H}{(1 - \alpha)J\mathbf{h}_{e,k}\mathbf{N_H}\mathbf{N_H}^H\mathbf{h}_{e,k}^H}\right),
$$

where $\tilde{\mathbf{f}}_j = \mathbf{f}_j/\|\mathbf{f}_j\|$ is the beamforming direction for the message intended for destination j. The lower bound is obtained by neglecting the contribution from the receiver noise. Since the second term does not vary with respect to $\bar{P}$, this shows that a positive secrecy rate can always be achieved by the AN-assisted beamforming scheme with sufficiently high transmit power, regardless of the number of eavesdroppers. This scheme is achieved without knowledge of the eavesdropper's channels at the source.

More generally, as mentioned previously, AN need not lie in the null space of the channels associated with all destinations and the beamforming vectors need not be orthogonal to each other. Specifically, similar to (3.2.2) and [15], let us consider the case where the AN covariance matrix $\mathbf{K_a}$ and the beamforming vectors $\mathbf{f}_j$, for $j = 1, \ldots, J$, are chosen to minimize the total transmit power subject to secrecy-rate constraints at the destinations. By assuming knowledge of both the destinations' and eavesdroppers' channel, the secrecy-rate-constrained power-minimization problem [15] can be formulated as follows:

$$
\min_{\mathbf{K_a},\mathbf{K}_j,\forall j} \sum_{j=1}^{J} \mathrm{tr}(\mathbf{K}_j) + \mathrm{tr}(\mathbf{K_a})
$$

$$
\text{subject to } \frac{1 + \dfrac{\mathbf{h}_{d,j}\mathbf{K}_j\mathbf{h}_{d,j}^H}{1 + \mathbf{h}_{d,j}\mathbf{K_a}\mathbf{h}_{d,j}^H + \sum_{\ell \neq j}\mathbf{h}_{d,j}\mathbf{K}_\ell\mathbf{h}_{d,j}^H}}{1 + \max\limits_k \dfrac{\mathbf{h}_{e,k}\mathbf{K}_j\mathbf{h}_{e,k}^H}{1 + \mathbf{h}_{e,k}\mathbf{K_a}\mathbf{h}_{e,k}^H + \sum_{\ell \neq j}\mathbf{h}_{e,k}\mathbf{K}_\ell\mathbf{h}_{e,k}^H}} \geq 2^{R_{0,j}}
$$

$$
\mathbf{K_a} \succeq \mathbf{0}, \mathbf{K}_j \succeq \mathbf{0}, \; \mathrm{rank}(\mathbf{K}_j) = 1, \; \forall j.
$$

By introducing auxiliary variables $\alpha_1, \ldots, \alpha_J$, the problem can be equivalently formulated as

$$
\min_{\mathbf{K_a},\mathbf{K}_j,\alpha_j \forall j} \sum_{j=1}^{J} \mathrm{tr}(\mathbf{K}_j) + \mathrm{tr}(\mathbf{K_a})
$$

$$\text{subject to } \mathbf{h}_{d,j}\mathbf{K}_j\mathbf{h}_{d,j}^H - (\alpha_j 2^{R_{0,j}}-1)\left(\sum_{\ell\neq j}\mathbf{h}_{d,j}\mathbf{K}_\ell\mathbf{h}_{d,j}^H + \mathbf{h}_{d,j}\mathbf{K_a}\mathbf{h}_{d,j}^H\right) \geq \alpha_j 2^{R_{0,j}}-1,$$

$$\mathbf{h}_{e,k}\mathbf{K}_j\mathbf{h}_{e,k}^H - (\alpha_j-1)\left(\sum_{\ell\neq j}\mathbf{h}_{e,k}\mathbf{K}_\ell\mathbf{h}_{e,k}^H + \mathbf{h}_{d,j}\mathbf{K_a}\mathbf{h}_{d,j}^H\right) \leq \alpha_j-1, \ \forall k,$$

$$\mathbf{K_a} \succeq 0, \ \mathbf{K}_j \succeq 0, \ \text{rank}(\mathbf{K}_j) = 1, \ \alpha_j \geq 0, \ \forall j.$$

Similar to (3.47a), this problem is still non-convex due to the rank-1 constraint. However, by relaxing the rank-1 constraint, the problem can then be efficiently solved using SDP for fixed choices of $\alpha_1, \ldots, \alpha_J$. Then, the variables $\alpha_1, \ldots, \alpha_J$ can be obtained using efficient gradient-based methods [15]. The solution obtained through this procedure may not be rank-1 and, thus, randomization methods [17] are also needed to extract from it the desired rank-1 solution.

3.5 Summary and Discussions

In this chapter, secrecy beamforming and precoding schemes were first introduced for the basic wiretap channel that consists of a source, a destination, and an eavesdropper. Beamforming refers to rank-1 transmissions by which only one data stream is sent over the multiple antennas whereas precoding refers to multi-rank transmissions by which more than one data stream can be transmitted simultaneously. In the case of secrecy beamforming, the optimal secrecy beamformer that maximizes the achievable secrecy rate, for the case where both the destination's and the eavesdropper's channels are available at the source, can be found explicitly and is given by a generalized eigenvector solution. In the case of secrecy precoding (with multiple antennas at the destination), the optimal secrecy precoder can only be found numerically under the sum power constraint, but can be obtained explicitly under the power-covariance constraint. The optimal precoder effectively decomposes the channel into multiple independent subchannels and utilizes only those that yield more favorable channel conditions to the destination than to the eavesdropper.

In addition to transmitting only information-bearing signals, the use of AN to degrade the eavesdropper's reception during secrecy transmission was also considered. This scheme also effectively increases the achievable secrecy rate, especially when the eavesdropper's channel is not completely known or when multiple destinations and eavesdroppers exist in the system. To prevent interference at the destination, AN is often embedded in the null space of the destination's channel. However, with partial knowledge of the eavesdropper's channel, it is sometimes desirable to shift AN towards the direction of the eavesdropper to cause stronger degradation at the eavesdropper even though this may cause AN leakage into the main channel [36].

It is worthwhile to remark that most secrecy beamforming/precoding and AN schemes introduced in this chapter were based on the secrecy-rate maximization criterion. In practice, the secrecy rate, which is derived under the perfect secrecy requirement and using random coding arguments, may often be over restrictive. Hence, designs based on other criteria, such as secrecy outage or signal-to-interference-plus-noise ratio (SINR), have also been considered in the literature. In particular, secrecy beamforming and AN designs have been examined in [37–39] based on the secrecy outage criterion and have been examined in [36, 40] based on the SINR criterion. Readers are referred to these works for further details on these designs.

References

1. Shafiee S, Liu N, Ulukus S (2009) Towards the secrecy capacity of the Gaussian MIMO wiretap channel: the 2–2-1 channel. IEEE Trans Inf Theory 55:4033–4039
2. Khisti A, Wornell G (2010) Secure transmission with multiple antennas I: the MISOME wiretap channel. IEEE Trans Inf Theory 56(7):3088–3104
3. Khisti A, Wornell G (2010) Secure transmission with multiple antennas. II. The MIMOME wiretap channel. IEEE Trans Inf Theory 56(11):5515–5532
4. Oggier F, Hassibi B (2011) The secrecy capacity of the MIMO wiretap channel. IEEE Trans Inf Theory 57(8):4961–4972
5. Harville DA (2008) Matrix algebra from a statisticians perspective. Springer, Berlin
6. Golub G, Loan CFV (1996) Matrix computations, 3rd edn. The Johns Hopkins Univeristy Press, Baltimore
7. Paige CC, Saunders MA (Jun. 1981) Towards a generalized singular value decomposition. SIAM J Numer Anal 18(3):389–405
8. Van Loan CF (Mar. 1976) Generalizing the singular value decomposition. SIAM J Numer Anal 13(1):76–83
9. Li J, Petropulu AP (2011) On beamforming solution for secrecy capacity of MIMO wiretap channels. In: Proceedings of IEEE Global Communications Conference (GLOBECOM) workshops, pp. 889–892
10. Bustin R, Liu R, Poor HV, Shamai (Shitz) S (2009) An MMSE approach to the secrecy capacity of the MIMO Gaussian wiretap channel. EURASIP J Wireless Commun Netw
11. Liang Y, Kramer G, Poor HV, Shamai (Shitz) S (2009) Compound wiretap channels. EURASIP J Wireless Commun Netw
12. Li Q, Ma W-K (2011) Optimal and robust transmit designs for MISO channel secrecy by semidefinite programming. IEEE Trans Signal Processing 59(8):3799–3812
13. Li Q, Ma W-K (2011) Multicast secrecy rate maximization for MISO channels with multiple multi-antenna eavesdroppers. In: Proceedings of IEEE International Conference on Communications (ICC)
14. Lei J, Han Z, Vázquez-Castro MA, Hjørungnes A (2011) Secure satellite communication systems design with individual secrecy rate constraints. IEEE Trans Inf Forensics Secur 6(3):661–671
15. Zheng G, Arapoglou P-D, Ottersten B (2012) Physical layer security in multibeam satellite systems. IEEE Trans Wireless Commun 11(2):852–862
16. Hamming RW (1962) Numerical Methods for Scientists and Engineers. McGraw-Hill, New York
17. Luo Z-Q, Ma W-K, So AM-C, Ye Y, Zhang S (May 2010) Semidefinite relaxation of quadratic optimization problems. IEEE Signal Processing Mag 27(3):20–34
18. Charnes A, Cooper WW (1962) Programming with linear fractional functionals. Naval Res Logistics Quart 9:181–186

19. Sturm JF (1999) Using SeDuMi 1.02, a MATLAB toolbox for optimization over symmetric cones. Optim Methods Softw 11–12:625–653. Version 1.05 available from http://fewcal.kub.nl/sturm

20. CVX Research, Inc. CVX: Matlab software for disciplined convex programming, version 2.0 beta, September 2012. Available http://cvxr.com/cvx

21. Sidiropoulos ND, Davidson TN, Luo Z-Q (2006) Transmit beamforming for physical-layer multicasting. IEEE Trans Signal Processing 54(6):2239–2251

22. Csiszàr I, Körner J (1978) Broadcast channels with confidential messages. IEEE Trans Inf Theory 24(3):339–348

23. Liu R, Maric I, Spasojević P, Yates RD (2008) Discrete memoryless interference and broadcast channels with confidential messages—secrecy rate regions. IEEE Trans Inf Theory 54(6):1–14

24. Liu R, Poor HV (2009) Secrecy capacity region of a multiple-antenna Gaussian broadcast channel with confidential messages. IEEE Trans Inf Theory 55(3):1235–1249

25. Tekin E, Yener A (2008) The Gaussian multiple access wire-tap channel. IEEE Trans Inf Theory 54(12):5747–5755

26. Liang Y, Poor HV (2008) Multiple-access channels with confidential messages. IEEE Trans Inf Theory 54(3):976–1002

27. Fakoorian SAA, Swindlehurst AL (2010) MIMO interference channel with confidential messages: game theoretic beamforming designs. In: Proceedings of 44th Signals, Systems, and Computers Conference, pp. 2099–2103

28. Geraci G, Egan M, Yuan J, Razi A, Collings IB (2012) Secrecy sum-rates for multi-user MIMO regularized channel inversion precoding. IEEE Trans Commun 60(11):3472–3482

29. Goel S, Negi R (2008) Guaranteeing secrecy using artificial noise. IEEE Trans Wireless Commun 7(6):2180–2189

30. Negi R, Goel S (2005) Secret communication using artificial noise. In: Proceedings of vehicular technology conference (VTC), vol 3, pp 1906–1910

31. Zhou X, McKay MR (2010) Secure transmission with artificial noise over fading channels: achievable rate and optimal power allocation. IEEE Trans Veh Technol 59(8):3831–3842

32. Gao H, Smith PJ, Clark MV (1998) Theoretical reliability of MMSE linear diversity combining in Rayleigh-fading additive interference channels. IEEE Trans Commun 46(5):666–672

33. Gradshteyn IS, Ryzhik IM (2007) Tables of integrals, series, and products, 7th edn. Dover, New York

34. Abramowitz M, Stegun IA (1974) Handbook of mathematical functions with formulaes, graphs, and mathematical tables. Dover, New York

35. Lin S-C, Chang T-H, Liang Y-L, Hong Y-WP, Chi C-Y (2011) On the impact of quantized channel feedback in guaranteeing secrecy with artificial noise: the noise leakage problem. IEEE Trans Wireless Commun 10(3):901–915

36. Liao W-C, Chang T-H, Ma W-K, Chi C-Y (2011) QoS-based transmit beamforming in the presence of eavesdroppers: an optimized artificial-noise-aided approach. IEEE Trans Signal Processing 59(3):1202–1216

37. Gerbracht S, Scheunert C, Jorswieck EA (2012) Secrecy outage in MISO systems with partial channel information. IEEE Trans Inf Forensics Secur 7(2):704–716

38. Xiong J, Wong K-K, Ma D, Wei J (2012) A closed-form power allocation for minimizing secrecy outage probability for MISO wiretap channels via masked beamforming. IEEE Commun Lett 16(9):1496–1499

39. Romero-Zurita N, Ghogho M, McLernon D (2012) Outage probability based power distribution between data and artificial noise for physical layer security. IEEE Signal Processing Lett 19(2):71–74

40. Mukherjee A, Swindlehurst AL (2011) Robust beamforming for security in MIMO wiretap channels with imperfect CSI. IEEE Trans Signal Processing 59(1):351–361

Chapter 4
Distributed Secrecy Beamforming and Precoding in Multi-Antenna Wireless Relay Systems

Abstract This chapter describes the use of the secrecy beamforming and precoding techniques in multiantenna wireless relay systems. Here, the multiple antennas at the relays may provide additional spatial degrees of freedom that can be further utilized to enhance secrecy. However, the extra source-to-relay transmission and the trustworthiness of the relays, i.e., whether or not the relays may also eavesdrop on the message, pose additional challenges in guaranteeing secrecy in these systems. The use of jamming or artificial noise (AN) in these scenarios is also discussed.

Keywords Relay · Distributed antenna system · Beamforming · Precoding · Artificial noise · Jamming · Secrecy.

Secrecy beamforming and precoding techniques, as introduced in Chap. 3, exploit the spatial degrees of freedom in multi-antenna systems to enlarge the channel quality difference between the destination and the eavesdropper. Interestingly, relay (or distributed antenna) systems can also be utilized by the source and the destination to provide the desired spatial degrees of freedom, even when each terminal is equipped with only a single antenna. The use of relaying to increase throughput and reliability in wireless systems has been studied extensively for conventional non-secrecy applications [1]. In secrecy applications, relays can be used not only to strengthen the signal at the destination but also to emit AN or jamming signals to disrupt the reception at the eavesdropper. However, with an additional node participating in the transmission, one should be concerned with the potential information leakage caused by the additional source-to-relay transmission and by the trustworthiness of the relay(s). These issues must be addressed in the design of distributed secrecy and precoding schemes in wireless relay systems.

Since the relay channel capacity and the optimal relaying strategy are not known in general, we consider in this chapter two types of relay strategies that have been adopted the most in the literature, namely, the decode-and-forward (DF) and the amplify-and-forward (AF) strategies, and utilize their achievable rates as the design criterion. Other types of relay strategies, e.g., the compress-and-forward (CF) strategy [2], have also been considered for the study of relay wiretap channels but will not be

Y.-W. P. Hong et al., *Signal Processing Approaches to Secure Physical Layer Communications in Multi-Antenna Wireless Systems*, SpringerBriefs in Signal Processing, DOI: 10.1007/978-981-4560-14-6_4, © The Author(s) 2014

discussed here. Furthermore, due to practical considerations, we also assume that all terminals are half-duplex and that two-phase transmissions are required in all relay strategies. In the following sections, secrecy beamforming and precoding schemes will be described for both trusted and untrusted relay systems.

4.1 Distributed Secrecy Beamforming and Precoding with Trusted Relays

In this section, distributed secrecy beamforming and precoding schemes are discussed for systems with trusted relays. Two relay strategies are considered, namely, the DF and the AF strategies.

Let us consider a basic relay wiretap channel that consists of a source, a destination, an eavesdropper, and a trusted relay (or relays), as illustrated in Fig. 4.1. The number of antennas at the source, the destination, the eavesdropper, and the relay are denoted by n_s, n_d, n_e, and n_r, respectively. The n_r relay antennas can be collocated at a single relay (as illustrated with the gray box with dashed line) or distributed among multiple relays (as illustrated with the smaller boxes with solid line). In both cases, the relay or relays can act as a distributed multi-antenna extension to the source and can be used to perform the desired secrecy beamforming and precoding. When the n_r antennas are collocated at a single relay, the signals received over the n_r antennas can be jointly processed and the signal transmitted over the relay antennas can be designed under a total relay power constraint. When the n_r antennas are distributed among multiple relays, the signal received at each relay can only be processed locally and the transmit signal should be designed under individual power constraints. In the following sections, we shall consider mostly the case where all n_r antennas are collocated at a single relay, but will also extend our discussions to the case with multiple relays in certain scenarios.

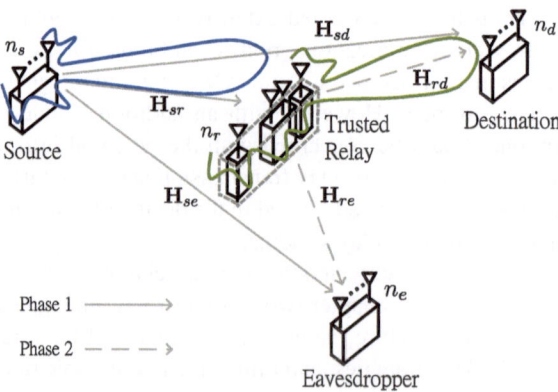

Fig. 4.1 Illustration of source and relay secrecy beamforming system

Specifically, relay transmissions are typically divided into two phases due to the half-duplex constraint. In Phase 1 (as illustrated by the solid arrows in Fig. 4.1), only the source transmits while the relay, the destination, and the eavesdropper receive. The signal transmitted by the source is denoted by the vector $\mathbf{x}_s \in \mathbb{C}^{n_s \times 1}$ and the signals received at the relay, the destination, and the eavesdropper are given respectively by

$$\mathbf{y}_r^{(1)} = \mathbf{H}_{sr}\mathbf{x}_s + \mathbf{w}_r^{(1)}, \tag{4.1a}$$

$$\mathbf{y}_d^{(1)} = \mathbf{H}_{sd}\mathbf{x}_s + \mathbf{w}_d^{(1)}, \tag{4.1b}$$

$$\mathbf{y}_e^{(1)} = \mathbf{H}_{se}\mathbf{x}_s + \mathbf{w}_e^{(1)}, \tag{4.1c}$$

where $\mathbf{H}_{sr} \in \mathbb{C}^{n_r \times n_s}$, $\mathbf{H}_{sd} \in \mathbb{C}^{n_d \times n_s}$, and $\mathbf{H}_{se} \in \mathbb{C}^{n_e \times n_s}$ are the source-to-relay, the source-to-destination, and the source-to-eavesdropper channels, and $\mathbf{w}_d^{(1)} \in \mathbb{C}^{n_d \times 1}$, $\mathbf{w}_r^{(1)} \in \mathbb{C}^{n_r \times 1}$, and $\mathbf{w}_e^{(1)} \in \mathbb{C}^{n_e \times 1}$ are independent additive white Gaussian noise (AWGN) vectors whose entries are i.i.d. $\mathcal{CN}(0, 1)$.

In Phase 2 (as illustrated by the dashed lines in Fig. 4.1), the relay forwards the signal $\mathbf{x}_r$ which, in general, can be viewed as a function of its received signal in Phase 1, i.e., $\mathbf{y}_r^{(1)}$. The received signals at the destination and the eavesdropper are expressed respectively as

$$\mathbf{y}_d^{(2)} = \mathbf{H}_{rd}\mathbf{x}_r + \mathbf{w}_d^{(2)}, \tag{4.2a}$$

$$\mathbf{y}_e^{(2)} = \mathbf{H}_{re}\mathbf{x}_r + \mathbf{w}_e^{(2)}, \tag{4.2b}$$

where $\mathbf{H}_{rd} \in \mathbb{C}^{n_d \times n_r}$ and $\mathbf{H}_{re} \in \mathbb{C}^{n_e \times n_r}$ are the relay-to-destination and the relay-to-eavesdropper channels, and $\mathbf{w}_d^{(2)}$ and $\mathbf{w}_e^{(2)}$ are independent AWGN vectors with entries that are i.i.d. $\mathcal{CN}(0, 1)$. The signal $\mathbf{x}_r$ transmitted by the relay depends on the specific type of relaying scheme employed and is given in more detail in the following sections.

4.1.1 Distributed Secrecy Beamforming and Precoding with Trusted Decode-and-Forward Relays

Let us consider a DF strategy where the relay first decodes the source message and then forwards it to the destination using the same codebook. Specifically, let $\mathbf{x}_s = \mathbf{F}_s\mathbf{u}$ be the signal transmitted by the source, where $\mathbf{F}_s \in \mathbb{C}^{n_s \times k}$ is the source precoder, $\mathbf{u} \in \mathcal{CN}(\mathbf{0}, \mathbf{I}_k)$ is the encoded symbol vector, and k is the number of signal dimensions. Given that the relay successfully decodes the message in Phase 1 and forwards the same codeword to the destination, the signal that it transmits in Phase 2 can be written as $\mathbf{x}_r = \mathbf{F}_r\mathbf{u}$, where $\mathbf{F}_r \in \mathbb{C}^{n_r \times k}$ is the relay precoder. In this case, the signal received at the destination and the eavesdropper over the two phases can be written as

$$\mathbf{y}_d = \begin{bmatrix} \mathbf{y}_d^{(1)} \\ \mathbf{y}_d^{(2)} \end{bmatrix} = \begin{bmatrix} \mathbf{H}_{sd}\mathbf{F}_s \\ \mathbf{H}_{rd}\mathbf{F}_r \end{bmatrix} \mathbf{u} + \begin{bmatrix} \mathbf{w}_d^{(1)} \\ \mathbf{w}_d^{(2)} \end{bmatrix} = \mathbf{G}_d\mathbf{u} + \mathbf{w}_d, \tag{4.3}$$

where $\mathbf{G}_d \triangleq [(\mathbf{H}_{sd}\mathbf{F}_s)^T \ (\mathbf{H}_{rd}\mathbf{F}_r)^T]^T$ and $\mathbf{w}_d \triangleq [(\mathbf{w}_d^{(1)})^T \ (\mathbf{w}_d^{(2)})^T]^T$, and

$$\mathbf{y}_e = \begin{bmatrix} \mathbf{y}_e^{(1)} \\ \mathbf{y}_e^{(2)} \end{bmatrix} = \begin{bmatrix} \mathbf{H}_{se}\mathbf{F}_s \\ \mathbf{H}_{re}\mathbf{F}_r \end{bmatrix} \mathbf{u} + \begin{bmatrix} \mathbf{w}_e^{(1)} \\ \mathbf{w}_e^{(2)} \end{bmatrix} = \mathbf{G}_e\mathbf{u} + \mathbf{w}_e, \tag{4.4}$$

where $\mathbf{G}_e \triangleq [(\mathbf{H}_{se}\mathbf{F}_s)^T \ (\mathbf{H}_{re}\mathbf{F}_r)^T]^T$ and $\mathbf{w}_e \triangleq [(\mathbf{w}_e^{(1)})^T \ (\mathbf{w}_e^{(2)})^T]^T$. Notice that, to employ the same codebook at the source and the relay, the signal dimension k must be less than both n_s and n_r (i.e., $k \leq \min\{n_s, n_r\}$). The transmit power at the source and the relay are given by $P_s = \mathrm{tr}(\mathbf{F}_s\mathbf{F}_s^H)$ and $P_r = \mathrm{tr}(\mathbf{F}_r\mathbf{F}_r^H)$.

Given $\mathbf{F}_s$, $\mathbf{F}_r$, and that the relay successfully decodes the confidential message, the achievable secrecy rate is given by

$$R_{\mathrm{DF}}^{(i)}(\mathbf{F}_s, \mathbf{F}_r) = \frac{1}{2} \left[I(\mathbf{x}_s, \mathbf{x}_r; \mathbf{y}_d) - I(\mathbf{x}_s, \mathbf{x}_r; \mathbf{y}_e) \right]^+, \tag{4.5}$$

where

$$\begin{aligned}
I(\mathbf{x}_s, \mathbf{x}_r; \mathbf{y}_d) &= h(\mathbf{y}_d) - h(\mathbf{y}_d|\mathbf{x}_s, \mathbf{x}_r) \\
&= \log_2 \det(\mathbf{I}_{2n_d} + \mathbf{G}_d\mathbf{G}_d^H) \\
&= \log_2 \det(\mathbf{I}_k + \mathbf{G}_d^H\mathbf{G}_d) \\
&= \log_2 \det(\mathbf{I}_k + \mathbf{F}_s^H\mathbf{H}_{sd}^H\mathbf{H}_{sd}\mathbf{F}_s + \mathbf{F}_r^H\mathbf{H}_{rd}^H\mathbf{H}_{rd}\mathbf{F}_r),
\end{aligned} \tag{4.6}$$

and

$$\begin{aligned}
I(\mathbf{x}_s, \mathbf{x}_r; \mathbf{y}_e) &= h(\mathbf{y}_e) - h(\mathbf{y}_e|\mathbf{x}_s, \mathbf{x}_r) \\
&= \log_2 \det(\mathbf{I}_{2n_e} + \mathbf{G}_e\mathbf{G}_e^H) \\
&= \log_2 \det(\mathbf{I}_k + \mathbf{F}_s^H\mathbf{H}_{se}^H\mathbf{H}_{se}\mathbf{F}_s + \mathbf{F}_r^H\mathbf{H}_{re}^H\mathbf{H}_{re}\mathbf{F}_r).
\end{aligned} \tag{4.7}$$

The factor $1/2$ appears in (4.5) due to the fact that two time units are used to transmit the same codeword (namely, in Phase 1 and Phase 2). This secrecy rate is achievable when the relay is able to successfully decode the message. However, to guarantee successful decoding at the relay (while maintaining confidentiality against the eavesdropper), the rate of the codeword must be less than

$$R_{\mathrm{DF}}^{(ii)}(\mathbf{F}_s, \mathbf{F}_r) = \frac{1}{2} \left[I(\mathbf{x}_s; \mathbf{y}_r) - I(\mathbf{x}_s, \mathbf{x}_r; \mathbf{y}_e) \right]^+, \tag{4.8}$$

where

$$I(\mathbf{x}_s; \mathbf{y}_r) = h(\mathbf{y}_r) - h(\mathbf{y}_r|\mathbf{x}_s)$$
$$= \log_2 \det(\mathbf{I}_{n_r} + \mathbf{H}_{sr}\mathbf{F}_s\mathbf{F}_s^H\mathbf{H}_{sr}^H)$$
$$= \log_2 \det(\mathbf{I}_k + \mathbf{F}_s^H\mathbf{H}_{sr}^H\mathbf{H}_{sr}\mathbf{F}_s) \tag{4.9}$$

and $I(\mathbf{x}_s, \mathbf{x}_r; \mathbf{y}_e)$ is given as in (4.7). Hence, given $\mathbf{F}_s$ and $\mathbf{F}_r$, the achievable secrecy rate for the DF strategy can be written as

$$R_{DF}(\mathbf{F}_s, \mathbf{F}_r) = \min\left\{ R_{DF}^{(i)}(\mathbf{F}_s, \mathbf{F}_r), R_{DF}^{(ii)}(\mathbf{F}_s, \mathbf{F}_r) \right\} \tag{4.10}$$

$$= \frac{1}{2}\left[\min\{I(\mathbf{x}_s, \mathbf{x}_r; \mathbf{y}_d), I(\mathbf{x}_s; \mathbf{y}_r)\} - I(\mathbf{x}_s, \mathbf{x}_r; \mathbf{y}_e)\right]^+ \tag{4.11}$$

$$= \frac{1}{2}\left[\min\left\{ \log_2 \frac{\det(\mathbf{I}_k + \mathbf{F}_s^H\mathbf{H}_{sd}^H\mathbf{H}_{sd}\mathbf{F}_s + \mathbf{F}_r^H\mathbf{H}_{rd}^H\mathbf{H}_{rd}\mathbf{F}_r)}{\det(\mathbf{I}_k + \mathbf{F}_s^H\mathbf{H}_{se}^H\mathbf{H}_{se}\mathbf{F}_s + \mathbf{F}_r^H\mathbf{H}_{re}^H\mathbf{H}_{re}\mathbf{F}_r)}, \right.$$
$$\left. \log_2 \frac{\det(\mathbf{I}_k + \mathbf{F}_s^H\mathbf{H}_{sr}^H\mathbf{H}_{sr}\mathbf{F}_s)}{\det(\mathbf{I}_k + \mathbf{F}_s^H\mathbf{H}_{se}^H\mathbf{H}_{se}\mathbf{F}_s + \mathbf{F}_r^H\mathbf{H}_{re}^H\mathbf{H}_{re}\mathbf{F}_r)} \right\}\right]^+. \tag{4.12}$$

The source and relay precoders can then be found by maximizing the achievable secrecy rate in (4.12). However, this problem is difficult to solve for several reasons. First of all, notice that the achievable secrecy rate for this case is given by minimum of two log-determinant functions. The secrecy precoder that maximizes one term may lead to smaller values for the other term, causing the latter to dominate the achievable secrecy rate. Second, finding the precoders to maximize each term in the minimization requires solving an optimization problem similar to that in Sect. 3.1.2. Even in those cases (i.e., the cases without relays), this problem is already considerably difficult and cannot be solved in closed-form. In the following, we discuss in more detail solutions for certain special cases previously considered in [3, 4].

Remark 1 : *It is worthwhile to remark that, in the DF scenario, the relay can alternatively choose to forward the source message using a codebook that is different from the source. In this case, the signals transmitted by the source and the relay can be written as* $\mathbf{x}_s = \mathbf{F}_s\mathbf{u}_s$ *and* $\mathbf{x}_r = \mathbf{F}_r\mathbf{u}_r$, *where* $\mathbf{u}_s$ *and* $\mathbf{u}_r$ *are the encoded symbol vectors with* k_s *and* k_r *signal dimensions, respectively. By assuming that the codebook at the source and the relay are independent, the achievable secrecy rate can be computed as*

$$R_{DF}(\mathbf{F}_s, \mathbf{F}_r)$$
$$= \frac{1}{2}\left[\min\left\{ \log_2 \frac{\det\left[(\mathbf{I}_{n_d} + \mathbf{H}_{sd}\mathbf{F}_s\mathbf{F}_s^H\mathbf{H}_{sd}^H)(\mathbf{I}_{n_d} + \mathbf{H}_{rd}\mathbf{F}_r\mathbf{F}_r^H\mathbf{H}_{rd}^H)\right]}{\det\left[(\mathbf{I}_{n_e} + \mathbf{H}_{se}\mathbf{F}_s\mathbf{F}_s^H\mathbf{H}_{se}^H)(\mathbf{I}_{n_e} + \mathbf{H}_{re}\mathbf{F}_r\mathbf{F}_r^H\mathbf{H}_{re}^H)\right]}, \right.$$
$$\left. \log_2 \frac{\det(\mathbf{I}_{n_r} + \mathbf{H}_{sr}\mathbf{F}_s\mathbf{F}_s^H\mathbf{H}_{sr}^H)}{\det\left[(\mathbf{I}_{n_e} + \mathbf{H}_{se}\mathbf{F}_s\mathbf{F}_s^H\mathbf{H}_{se}^H)(\mathbf{I}_{n_e} + \mathbf{H}_{re}\mathbf{F}_r\mathbf{F}_r^H\mathbf{H}_{re}^H)\right]} \right\}\right]^+. \tag{4.13}$$

The source and relay precoders $\mathbf{F}_s$ and $\mathbf{F}_r$ can then be chosen to maximize the achievable secrecy rate in (4.13).

Specifically, let us consider the special case where the source, the destination, and the eavesdropper are all equipped with only a single antenna (i.e., $n_s = n_d = n_e = 1$) whereas the relay is assumed to have multiple antennas (i.e., $n_r > 1$). In this case, the channel matrices $\mathbf{H}_{sr}$, $\mathbf{H}_{rd}$ and $\mathbf{H}_{re}$ can be replaced with column vector $\mathbf{h}_{sr} \in \mathbb{C}^{n_r \times 1}$ and row vectors $\mathbf{h}_{rd} \in \mathbb{C}^{1 \times n_r}$ and $\mathbf{h}_{re} \in \mathbb{C}^{1 \times n_r}$, respectively. Similarly, the channel matrices $\mathbf{H}_{sd}$ and $\mathbf{H}_{se}$ can be replaced with scalars h_{sd} and h_{se}, respectively. With $n_s = 1$, the number of signal dimensions, i.e., k, is at most 1 and, thus, the signal transmitted by the source in Phase 1 can be written as $x_s = \sqrt{P_s}u$, where $u \in \mathcal{CN}(0, 1)$. By employing the same codebook at the relay, the signal transmitted by the relay in Phase 2 can be written as $\mathbf{x}_r = \mathbf{f}_r u$ with $||\mathbf{f}_r||^2 = P_r$. (Notice that $\mathbf{F}_s \in \mathbb{C}^{n_s \times k}$ is simply equal to $\sqrt{P_s}$, since $n_s = k = 1$, and $\mathbf{F}_r \in \mathbb{C}^{n_r \times k}$ is replaced with the column vector $\mathbf{f}_r \in \mathbb{C}^{n_r \times 1}$.) Following the formulation in (4.12), the achievable secrecy rate given $(P_s, \mathbf{f}_r)$ can be written as

$$R_{\text{DF}}(P_s, \mathbf{f}_r) = \frac{1}{2}\left[\min\left\{\log_2 \frac{1 + P_s|h_{sd}|^2 + \mathbf{f}_r^H \mathbf{h}_{rd}^H \mathbf{h}_{rd} \mathbf{f}_r}{1 + P_s|h_{se}|^2 + \mathbf{f}_r^H \mathbf{h}_{re}^H \mathbf{h}_{re} \mathbf{f}_r},\right.\right.$$
$$\left.\left.\log_2 \frac{1 + P_s\|\mathbf{h}_{sr}\|^2}{1 + P_s|h_{se}|^2 + \mathbf{f}_r^H \mathbf{h}_{re}^H \mathbf{h}_{re} \mathbf{f}_r}\right\}\right]^+ \tag{4.14}$$

Our goal is to choose P_s and $\mathbf{f}_r$ to maximize the achievable secrecy rate subject to the total power constraint $P_s + P_r = \bar{P}$.

Notice that, when $\|\mathbf{h}_{sr}\|^2 < |h_{sd}|^2$, we have

$$1 + P_s\|\mathbf{h}_{sr}\|^2 < 1 + P_s|h_{sd}|^2 + \mathbf{f}_r^H \mathbf{h}_{rd}^H \mathbf{h}_{rd} \mathbf{f}_r \tag{4.15}$$

for any P_s and $\mathbf{f}_r$ that are not simultaneously zero. In this case, the achievable secrecy rate is given by

$$R_{\text{DF}}(P_s, \mathbf{f}_r; \|\mathbf{h}_{sr}\|^2 < |h_{sd}|^2) = \frac{1}{2}\left[\log_2 \frac{1 + P_s\|\mathbf{h}_{sr}\|^2}{1 + P_s|h_{se}|^2 + \mathbf{f}_r^H \mathbf{h}_{re}^H \mathbf{h}_{re} \mathbf{f}_r}\right]^+ \tag{4.16}$$

and is maximized under the total power constraint $P_s + P_r = \bar{P}$ by choosing $P_s = \bar{P}$ and $\mathbf{f}_r = \mathbf{0}$ (i.e., $P_r = 0$).

On the other hand, when $\|\mathbf{h}_{sr}\|^2 \geq |h_{sd}|^2$, one can always find $(P_s, \mathbf{f}_r)$ such that

$$\frac{1 + P_s|h_{sd}|^2 + \mathbf{f}_r^H \mathbf{h}_{rd}^H \mathbf{h}_{rd} \mathbf{f}_r}{1 + P_s|h_{se}|^2 + \mathbf{f}_r^H \mathbf{h}_{re}^H \mathbf{h}_{re} \mathbf{f}_r} \leq \frac{1 + P_s\|\mathbf{h}_{sr}\|^2}{1 + P_s|h_{se}|^2 + \mathbf{f}_r^H \mathbf{h}_{re}^H \mathbf{h}_{re} \mathbf{f}_r}. \tag{4.17}$$

By defining $\tilde{\mathbf{f}}_r = \mathbf{f}_r/\sqrt{P_r}$ as the beamforming direction, the condition in (4.17) can be equivalently written as $P_r \tilde{\mathbf{f}}_r^H \mathbf{h}_{rd}^H \mathbf{h}_{rd} \tilde{\mathbf{f}}_r < P_s(\|\mathbf{h}_{sr}\|^2 - |h_{sd}|^2)$. Under the total power constraint $P_s + P_r = \bar{P}$, this yields the following condition on the source transmit power:

$$P_s \geq \bar{P} \frac{\tilde{\mathbf{f}}_r^H \mathbf{h}_{rd}^H \mathbf{h}_{rd} \tilde{\mathbf{f}}_r}{\tilde{\mathbf{f}}_r^H \left[(\|\mathbf{h}_{sr}\|^2 - |h_{sd}|^2)\mathbf{I} + \mathbf{h}_{rd}^H \mathbf{h}_{rd} \right] \tilde{\mathbf{f}}_r} \triangleq P_{s,\min}(\tilde{\mathbf{f}}_r). \qquad (4.18)$$

Here, $P_{s,\min}(\tilde{\mathbf{f}}_r)$ can be viewed as the minimum source transmit power required to ensure that the secrecy rate is not limited by the second term inside the minimization of (4.14) (i.e., the secrecy rate on the source-to-relay channel). It follows by the Rayleigh-Ritz Theorem [5] that $0 \leq \min_{\mathbf{f}:\|\mathbf{f}\|=1} P_{s,\min}(\tilde{\mathbf{f}}) \leq \bar{P}$ and, thus, there always exists P_s such that (4.18) holds. In this case, the achievable secrecy rate is given by

$$R_{\mathrm{DF}}(P_s, \mathbf{f}_r; \|\mathbf{h}_{sr}\|^2 \geq |h_{sd}|^2)$$

$$= \frac{1}{2} \left[\log_2 \frac{1 + P_s|h_{sd}|^2 + \mathbf{f}_r^H \mathbf{h}_{rd}^H \mathbf{h}_{rd} \mathbf{f}_r}{1 + P_s|h_{se}|^2 + \mathbf{f}_r^H \mathbf{h}_{re}^H \mathbf{h}_{re} \mathbf{f}_r} \right]^+ \qquad (4.19)$$

$$= \frac{1}{2} \left[\log_2 \frac{\tilde{\mathbf{f}}_r^H (\mathbf{I}_{n_r} + P_s|h_{sd}|^2 \mathbf{I}_{n_r} + P_r \mathbf{h}_{rd}^H \mathbf{h}_{rd}) \tilde{\mathbf{f}}_r}{\tilde{\mathbf{f}}_r^H (\mathbf{I}_{n_r} + P_s|h_{se}|^2 \mathbf{I}_{n_r} + P_r \mathbf{h}_{re}^H \mathbf{h}_{re}) \tilde{\mathbf{f}}_r} \right]^+. \qquad (4.20)$$

Due to the monotonicity of the logarithmic function, maximizing the achievable secrecy rate is equivalent to maximizing the ratio inside. Therefore, the optimization problem can be formulated as follows:

$$\max_{P_s, P_r, \tilde{\mathbf{f}}_r} \quad \frac{\tilde{\mathbf{f}}_r^H (\mathbf{I}_{n_r} + P_s|h_{sd}|^2 \mathbf{I}_{n_r} + P_r \mathbf{h}_{rd}^H \mathbf{h}_{rd}) \tilde{\mathbf{f}}_r}{\tilde{\mathbf{f}}_r^H (\mathbf{I}_{n_r} + P_s|h_{se}|^2 \mathbf{I}_{n_r} + P_r \mathbf{h}_{re}^H \mathbf{h}_{re}) \tilde{\mathbf{f}}_r} \qquad (4.21a)$$

$$\text{subject to} \quad P_{s,\min}(\tilde{\mathbf{f}}_r) \leq P_s \leq \bar{P}, \quad P_r = \bar{P} - P_s, \qquad (4.21b)$$

$$\text{and } \|\tilde{\mathbf{f}}_r\|^2 = 1. \qquad (4.21c)$$

This optimization problem is non-convex in general and is difficult to solve in closed-form. Though it is difficult to optimize over P_s, P_r, and $\tilde{\mathbf{f}}_r$ jointly, the optimal beamforming direction $\tilde{\mathbf{f}}_r$ can be solved separately for fixed values of P_s and P_r, and vice versa. Utilizing this property, an approximate solution for this problem can then be efficiently obtained by performing an alternate hill-climbing optimization method as described in the following.

Specifically, given P_s and P_r, the optimal beamforming direction can be solved following derivations similar to that in Sect. 3.1 and is given by

$$\tilde{\mathbf{f}}_r^* = \psi_{\max} \left(\mathbf{I}_{n_r} + P_s|h_{sd}|^2 \mathbf{I}_{n_r} + P_r \mathbf{h}_{rd}^H \mathbf{h}_{rd}, \mathbf{I}_{n_r} + P_s|h_{se}|^2 \mathbf{I}_{n_r} + P_r \mathbf{h}_{re}^H \mathbf{h}_{re} \right),$$
$$(4.22)$$

where $\psi_{\max}(\mathbf{A}, \mathbf{B})$ denotes the normalized generalized eigenvector corresponding to the maximum generalized eigenvalue of the matrix pair $(\mathbf{A}, \mathbf{B})$. The resulting achievable secrecy rate in this case is given by

$$
\begin{aligned}
&R_{\mathrm{DF}}^{(i)*}(P_s, P_r) \\
&= \frac{1}{2}\left[\log_2 \lambda_{\max}\left(\mathbf{I}_{n_r}+P_s|h_{sd}|^2\mathbf{I}_{n_r}+P_r\mathbf{h}_{rd}^H\mathbf{h}_{rd}, \mathbf{I}_{n_r}+P_s|h_{se}|^2\mathbf{I}_{n_r}+P_r\mathbf{h}_{re}^H\mathbf{h}_{re}\right)\right]^+,
\end{aligned}
\tag{4.23}
$$

where $\lambda_{\max}(\mathbf{A}, \mathbf{B})$ denotes the maximum generalized eigenvalue of the matrix pair $(\mathbf{A}, \mathbf{B})$. On the other hand, when $\tilde{\mathbf{f}}_r$ is fixed, the power allocation problem can be formulated as

$$
\max_{P_s} \quad \frac{1+P_s|h_{sd}|^2+(\bar{P}-P_s)|\mathbf{h}_{rd}\tilde{\mathbf{f}}_r|^2}{1+P_s|h_{se}|^2+(\bar{P}-P_s)|\mathbf{h}_{re}\tilde{\mathbf{f}}_r|^2}
\tag{4.24a}
$$

$$
\text{subject to} \quad P_{s,\min}(\tilde{\mathbf{f}}_r) \le P_s \le \bar{P}.
\tag{4.24b}
$$

By the fact that the function $f(x) = (a+bx)/(c+dx)$ increases monotonically with x when $bc > da$, the optimal source and relay powers are given by

$$
P_s^* =
\begin{cases}
\bar{P}, & \text{if } \dfrac{|h_{sd}|^2-|\mathbf{h}_{rd}\tilde{\mathbf{f}}_r|^2}{1+\bar{P}|\mathbf{h}_{rd}\tilde{\mathbf{f}}_r|^2} > \dfrac{|h_{se}|^2-|\mathbf{h}_{re}\tilde{\mathbf{f}}_r|^2}{1+\bar{P}|\mathbf{h}_{re}\tilde{\mathbf{f}}_r|^2} \\[4ex]
P_{s,\min}(\tilde{\mathbf{f}}_r), & \text{if } \dfrac{|h_{sd}|^2-|\mathbf{h}_{rd}\tilde{\mathbf{f}}_r|^2}{1+\bar{P}|\mathbf{h}_{rd}\tilde{\mathbf{f}}_r|^2} \le \dfrac{|h_{se}|^2-|\mathbf{h}_{re}\tilde{\mathbf{f}}_r|^2}{1+\bar{P}|\mathbf{h}_{re}\tilde{\mathbf{f}}_r|^2}
\end{cases}
\tag{4.25}
$$

and

$$
P_r^* = \bar{P} - P_s^*.
\tag{4.26}
$$

Based on the above results, the hill-climbing algorithm can then be employed as in [3] and as described below.

Hill-Climbing Algorithm

1. Choose an initial value of $\tilde{\mathbf{f}}_r$ such that $\|\tilde{\mathbf{f}}_r\|^2 = 1$ and $P_{s,\min}(\tilde{\mathbf{f}}_r) < \bar{P}$.
2. Compute the optimal source and relay powers P_s and P_r by (4.25) and (4.26), respectively, for fixed beamforming direction $\tilde{\mathbf{f}}_r$.
3. Compute the optimal beamforming direction $\tilde{\mathbf{f}}_r$ by (4.22) for fixed power values P_s and P_r.
4. Iterate Steps 2 and 3 until the solution converges.

Notice that the hill-climbing algorithm in general does not guarantee convergence to the optimal solution if the maximum generalized eigenvalue is not a convex function of P_s and P_r. If the globally optimal solution is desired, the conventional random search algorithm can be employed at the cost of considerably more iterations. The readers are referred to [3, 6] for further details on the random search algorithm. The above results can also be readily extended to the case with multiple antennas at the destination and the eavesdropper by replacing the channel vectors $\mathbf{h}_{rd}$ and $\mathbf{h}_{re}$ with the channel matrices $\mathbf{H}_{rd}$ and $\mathbf{H}_{re}$.

Alternatively, extending upon the previous special case, we can also consider the case where the n_r relay antennas are located separately at n_r relays, i.e., each relay is equipped with only a single antenna. Here, the relays are assumed to be able to adjust their gains and phases so as to collaboratively achieve the desired beamforming effect. Similar to before, the received signal at the destination and the eavesdropper can be written as (4.3) and (4.4) with $\mathbf{H}_{sr}$, $\mathbf{H}_{rd}$ and $\mathbf{H}_{re}$ replaced with $\mathbf{h}_{sr} = [h_{s,r_1}, \ldots, h_{s,r_{n_r}}]^T$, $\mathbf{h}_{rd} = [h_{r_1,d}, \ldots, h_{r_{n_r},d}]$, and $\mathbf{h}_{re} = [h_{r_1,e}, \ldots, h_{r_{n_r},e}]$, respectively, and with $\mathbf{H}_{sd}$ and $\mathbf{H}_{se}$ replaced with scalars h_{sd} and h_{se}, respectively. The entries h_{s,r_i}, $h_{r_i,d}$, and $h_{r_i,e}$ represent the channel from the source to relay i, from relay i to the destination, and from relay i to the eavesdropper, respectively. In this case, the relay beamforming vector can be written as $\mathbf{f}_r = [f_{r_1}, \ldots, f_{r_{n_r}}]^T$, where f_{r_i} is the relay gain imposed by the ith relay. Similar to (4.14), the achievable secrecy rate given $(P_s, \mathbf{f}_r)$ in this case can be written as

$$R_{DF}(P_s, \mathbf{f}_r) = \frac{1}{2} \left[\min \left\{ \log_2 \frac{1 + P_s|h_{sd}|^2 + \mathbf{f}_r^H \mathbf{h}_{rd}^H \mathbf{h}_{rd} \mathbf{f}_r}{1 + P_s|h_{se}|^2 + \mathbf{f}_r^H \mathbf{h}_{re}^H \mathbf{h}_{re} \mathbf{f}_r}, \right. \right.$$
$$\left. \left. \min_{i=1,\ldots,n_r} \log_2 \frac{1 + P_s|h_{s,r_i}|^2}{1 + P_s|h_{se}|^2 + \mathbf{f}_r^H \mathbf{h}_{re}^H \mathbf{h}_{re} \mathbf{f}_r} \right\} \right]^+ . \qquad (4.27)$$

The second term contains the minimization over the secrecy rate achievable between the source and the n_r relays. This represents the maximum secrecy rate that can be adopted to guarantee successful decoding at all n_r relays.

Suppose the source and relays are subject to a total power constraint $P_s + \sum_{i=1}^{n_r} P_{r_i} = \bar{P}$, where $P_{r_i} = |f_{r_i}|^2$ is relay i's transmit power. Similar to the collocated antenna case, if $|h_{s,r_i}|^2 < |h_{sd}|^2$ for some i, then it follows that

$$1 + P_s|h_{s,r_i}|^2 < 1 + P_s|h_{sd}|^2 + \mathbf{f}_r^H \mathbf{h}_{rd}^H \mathbf{h}_{rd} \mathbf{f}_r \qquad (4.28)$$

for any P_s and $\mathbf{f}_r$ that are not simultaneously zero. In this case, the achievable secrecy rate is given by

$$R_{DF}(P_s, \mathbf{f}_r; \exists\, i \text{ s.t. } |h_{s,r_i}|^2 < |h_{sd}|^2)$$
$$= \frac{1}{2} \left[\min_{i=1,\ldots,n_r} \log_2 \frac{1 + P_s|h_{s,r_i}|^2}{1 + P_s|h_{se}|^2 + \mathbf{f}_r^H \mathbf{h}_{re}^H \mathbf{h}_{re} \mathbf{f}_r} \right]^+ \qquad (4.29)$$

and is maximized under the total power constraint $P_s + \sum_{i=1}^{n_r} P_{r_i} = \bar{P}$ by choosing $P_s = \bar{P}$ and $\mathbf{f}_r = \mathbf{0}$ (i.e., $\sum_{i=1}^{n_r} P_{r_i} = 0$).

On the other hand, if $|h_{s,r_i}|^2 \geq |h_{sd}|^2$, $\forall i$, then there exists $(P_s, \mathbf{f}_r)$ such that

$$\frac{1 + P_s|h_{sd}|^2 + \mathbf{f}_r^H \mathbf{h}_{rd}^H \mathbf{h}_{rd} \mathbf{f}_r}{1 + P_s|h_{se}|^2 + \mathbf{f}_r^H \mathbf{h}_{re}^H \mathbf{h}_{re} \mathbf{f}_r} \leq \frac{1 + P_s|h_{s,r_i}|^2}{1 + P_s|h_{se}|^2 + \mathbf{f}_r^H \mathbf{h}_{re}^H \mathbf{h}_{re} \mathbf{f}_r}, \tag{4.30}$$

for all i. Under the total power constraint $P_s + \sum_{i=1}^{n_r} P_{r_i} = \bar{P}$, i.e., with $\sum_{i=1}^{n_r} P_{r_i} = \bar{P} - P_s$, this yields the conditions

$$P_s \geq \bar{P} \frac{\tilde{\mathbf{f}}_r^H \mathbf{h}_{rd}^H \mathbf{h}_{rd} \tilde{\mathbf{f}}_r}{\tilde{\mathbf{f}}_r \left[(|h_{sr_i}|^2 - |h_{sd}|^2)\mathbf{I} + \mathbf{h}_{rd}^H \mathbf{h}_{rd} \right] \tilde{\mathbf{f}}_r} \triangleq P_{s,\min}^{(i)}(\tilde{\mathbf{f}}_r) \tag{4.31}$$

for $i = 1, \ldots, n_r$. Similarly, $P_{s,\min}^{(i)}(\tilde{\mathbf{f}}_r)$ can be viewed as the minimum source transmit power required to ensure that the achievable secrecy rate is not limited by the channel between the source and relay i. In this case, the achievable secrecy rate is given by

$$R_{\mathrm{DF}}(P_s, \mathbf{f}_r; |h_{s,r_i}|^2 \geq |h_{sd}|^2, \forall i) = \frac{1}{2} \left[\log_2 \frac{\tilde{\mathbf{f}}_r^H (\mathbf{I}_{n_r} + P_s|h_{sd}|^2 \mathbf{I}_{n_r} + P_r \mathbf{h}_{rd}^H \mathbf{h}_{rd}) \tilde{\mathbf{f}}_r}{\tilde{\mathbf{f}}_r^H (\mathbf{I}_{n_r} + P_s|h_{se}|^2 \mathbf{I}_{n_r} + P_r \mathbf{h}_{re}^H \mathbf{h}_{re}) \tilde{\mathbf{f}}_r} \right]^+ \tag{4.32}$$

where $\tilde{\mathbf{f}}_r \triangleq \mathbf{f}_r / \sqrt{P_r}$ is the collaborative beamforming direction and $P_r = \sum_{i=1}^{n_r} P_{r_i}$ is the total relay power. In this case, the optimization problem can be formulated, similar to (4.21), as

$$\max_{P_s, P_r, \tilde{\mathbf{f}}_r} \quad \frac{\tilde{\mathbf{f}}_r^H (\mathbf{I}_{n_r} + P_s|h_{sd}|^2 \mathbf{I}_{n_r} + P_r \mathbf{h}_{rd}^H \mathbf{h}_{rd}) \tilde{\mathbf{f}}_r}{\tilde{\mathbf{f}}_r^H (\mathbf{I}_{n_r} + P_s|h_{se}|^2 \mathbf{I}_{n_r} + P_r \mathbf{h}_{re}^H \mathbf{h}_{re}) \tilde{\mathbf{f}}_r} \tag{4.33a}$$

$$\text{subject to} \quad \max_{i=1,\ldots,n_r} P_{s,\min}^{(i)}(\tilde{\mathbf{f}}_r) \leq P_s \leq \bar{P}, \quad P_r = \bar{P} - P_s, \tag{4.33b}$$

$$\text{and} \ \|\tilde{\mathbf{f}}_r\|^2 = 1. \tag{4.33c}$$

Similar to the previous case, for fixed P_s and P_r, the optimal collaborative beamforming direction is given by

$$\mathbf{f}_r^* = \sqrt{P_r} \psi_{\max} \left(\mathbf{I}_{n_r} + P_s|h_{sd}|^2 \mathbf{I}_{n_r} + P_r \mathbf{h}_{rd}^H \mathbf{h}_{rd}, \ \mathbf{I}_{n_r} + P_s|h_{se}|^2 \mathbf{I}_{n_r} + P_r \mathbf{h}_{re}^H \mathbf{h}_{re} \right) \tag{4.34}$$

and the resulting achievable secrecy rate is given by

$$R_{\mathrm{DF}}^{(i)*}(P_s, P_r)$$
$$= \frac{1}{2} \left[\log_2 \lambda_{\max} \left(\mathbf{I}_{n_r} + P_s|h_{sd}|^2 \mathbf{I}_{n_r} + P_r \mathbf{h}_{rd}^H \mathbf{h}_{rd}, \ \mathbf{I}_{n_r} + P_s|h_{se}|^2 \mathbf{I}_{n_r} + P_r \mathbf{h}_{re}^H \mathbf{h}_{re} \right) \right]^+. \tag{4.35}$$

On the other hand, for fixed $\tilde{\mathbf{f}}_r$, the optimal source and relay powers are given by

$$P_s^* = \begin{cases} \bar{P}, & \text{if } \dfrac{|h_{sd}|^2 - |\mathbf{h}_{rd}\tilde{\mathbf{f}}_r|^2}{1 + \bar{P}|\mathbf{h}_{rd}\tilde{\mathbf{f}}_r|^2} > \dfrac{|h_{se}|^2 - |\mathbf{h}_{re}\tilde{\mathbf{f}}_r|^2}{1 + \bar{P}|\mathbf{h}_{re}\tilde{\mathbf{f}}_r|^2} \\[3mm] \max\limits_{i=1,\ldots,n_r} P_{s,min}^{(i)}(\tilde{\mathbf{f}}_r), & \text{if } \dfrac{|h_{sd}|^2 - |\mathbf{h}_{rd}\tilde{\mathbf{f}}_r|^2}{1 + \bar{P}|\mathbf{h}_{rd}\tilde{\mathbf{f}}_r|^2} \le \dfrac{|h_{se}|^2 - |\mathbf{h}_{re}\tilde{\mathbf{f}}_r|^2}{1 + \bar{P}|\mathbf{h}_{re}\tilde{\mathbf{f}}_r|^2} \end{cases} \tag{4.36}$$

and

$$P_r^* = \bar{P} - P_s^*. \tag{4.37}$$

Based on the above results, a solution for the optimization problem in (4.33) can also be obtained (at least approximately) using the hill-climbing algorithm described previously.

Interestingly, one can observe from (4.27) that, in the case of separately located antennas, increasing the number of antennas (i.e., relays) may, on the one hand, increase the diversity advantages but may, on the other hand, limit the secrecy rate due to the need to successfully decode at all relays. Hence, intelligent relay selection policies can also be devised to further increase the achievable secrecy rate. In [3], the authors proposed a simple relay selection policy based on the values of $P_{s,min}^{(i)}$, for all i, as given in (4.31). Specifically, suppose the relay indices are chosen such that $P_{s,min}^{(1)}(\tilde{\mathbf{f}}_r) \le P_{s,min}^{(2)}(\tilde{\mathbf{f}}_r) \le \cdots \le P_{s,min}^{(n_r)}(\tilde{\mathbf{f}}_r)$. The relay selection policy proposed in [3] simply chooses the first J relays (i.e., the relays associated with the J smallest source power constraints $P_{s,min}^{(i)}$, for $i = 1, \ldots, J$) as the participating relays. This loosens the constraint on the source transmit power, but reduces the spatial degrees of freedom that can be utilized for beamforming in Phase 2. The optimum value of J is obtained via exhaustive search.

It is necessary to remark that, in the above, we considered only the total power constraint among relays. In practice, it may be more relevant to consider the case where each relay is subject to its own individual power constraint. This problem was examined in [4] and was solved using the semi-definite relaxation (SDR) approach.

4.1.2 Distributed Secrecy Beamforming and Precoding with Trusted Amplify-and-Forward Relays

In this section, we consider the design of secrecy beamforming and precoding schemes for trusted amplify-and-forward (AF) relays. Here, we also adopt a two-phase transmission strategy, where the relay first receives the signal from the source in Phase 1 and then forwards a linearly amplified version of the received signal to the destination in Phase 2.

Specifically, in Phase 1, the source transmits the signal $\mathbf{x}_s = \mathbf{F}_s\mathbf{u}$, where $\mathbf{F}_s \in \mathbb{C}^{n_s \times k}$ is the source precoder, $\mathbf{u} \in \mathcal{CN}(\mathbf{0}, \mathbf{I}_k)$ is the encoded symbol vector, and k is the number of signal dimensions. The source transmit power is given by $P_s = \mathrm{E}[\|\mathbf{x}_s\|^2] = \mathrm{tr}(\mathbf{F}_s\mathbf{F}_s^H)$, and the received signals at the relay, the destination, and the eavesdropper in Phase 1 are given as in (4.1). Then, in Phase 2, the relay forwards a linearly amplified version of its received signal $\mathbf{y}_r^{(1)}$ to the destination. The signal transmitted by the relay can be written as

$$\mathbf{x}_r = \mathbf{F}_r \mathbf{y}_r^{(1)} \tag{4.38}$$

where $\mathbf{F}_r \in \mathbb{C}^{n_s \times n_s}$ is the relay precoder. The relay transmit power is $P_r = E[\|\mathbf{x}_r\|^2] = \mathrm{tr}(\mathbf{F}_r(\mathbf{H}_{sr}\mathbf{F}_s\mathbf{F}_s^H\mathbf{H}_{sr}^H + \mathbf{I}_{n_r})\mathbf{F}_r^H)$. The received signals at the destination and the eavesdropper in Phase 2 are given as in (4.2). For convenience, we combine the received signals in both phases into the effective signal vectors

$$\mathbf{y}_d = \begin{bmatrix} \mathbf{y}_d^{(1)} \\ \mathbf{y}_d^{(2)} \end{bmatrix} = \underbrace{\begin{bmatrix} \mathbf{H}_{sd} \\ \mathbf{H}_{rd}\mathbf{F}_r\mathbf{H}_{sr} \end{bmatrix}}_{\triangleq \tilde{\mathbf{H}}_{sd}} \mathbf{x}_s + \underbrace{\begin{bmatrix} \mathbf{w}_d^{(1)} \\ \mathbf{H}_{rd}\mathbf{F}_r\mathbf{w}_r^{(1)} + \mathbf{w}_d^{(2)} \end{bmatrix}}_{\triangleq \tilde{\mathbf{w}}_d} \tag{4.39}$$

$$\triangleq \tilde{\mathbf{H}}_{sd}\mathbf{x}_s + \tilde{\mathbf{w}}_d \tag{4.40}$$

and

$$\mathbf{y}_e = \begin{bmatrix} \mathbf{y}_e^{(1)} \\ \mathbf{y}_e^{(2)} \end{bmatrix} = \underbrace{\begin{bmatrix} \mathbf{H}_{se} \\ \mathbf{H}_{re}\mathbf{F}_r\mathbf{H}_{sr} \end{bmatrix}}_{\triangleq \tilde{\mathbf{H}}_{se}} \mathbf{x}_s + \underbrace{\begin{bmatrix} \mathbf{w}_e^{(1)} \\ \mathbf{H}_{re}\mathbf{F}_r\mathbf{w}_r^{(1)} + \mathbf{w}_e^{(2)} \end{bmatrix}}_{\triangleq \tilde{\mathbf{w}}_e} \tag{4.41}$$

$$= \tilde{\mathbf{H}}_{se}\mathbf{x}_s + \tilde{\mathbf{w}}_e, \tag{4.42}$$

where $\tilde{\mathbf{H}}_{sd}$ and $\tilde{\mathbf{H}}_{se}$ are the effective channel matrices viewed by the destination and the eavesdropper, respectively, and $\tilde{\mathbf{w}}_d$ and $\tilde{\mathbf{w}}_e$ are the effective noise vectors, whose covariance matrices are given by

$$\mathbf{K}_{\tilde{\mathbf{w}}_d} = E[\tilde{\mathbf{w}}_d\tilde{\mathbf{w}}_d^H] = \begin{bmatrix} \mathbf{I}_{n_d} & \mathbf{0} \\ \mathbf{0} & \mathbf{I}_{n_d} + \mathbf{H}_{rd}\mathbf{F}_r\mathbf{F}_r^H\mathbf{H}_{rd}^H \end{bmatrix} \tag{4.43}$$

and

$$\mathbf{K}_{\tilde{\mathbf{w}}_e} = E[\tilde{\mathbf{w}}_e\tilde{\mathbf{w}}_e^H] = \begin{bmatrix} \mathbf{I}_{n_e} & \mathbf{0} \\ \mathbf{0} & \mathbf{I}_{n_e} + \mathbf{H}_{re}\mathbf{F}_r\mathbf{F}_r^H\mathbf{H}_{re}^H \end{bmatrix}. \tag{4.44}$$

By treating $\mathbf{x}_s$ as the channel input and $\mathbf{y}_d$ and $\mathbf{y}_e$ as the effective channel outputs at the destination and the eavesdropper, respectively, the AF system can be viewed effectively as a multiple-input multiple-output (MIMO) system, similar to that described in (2.16). Hence, the achievable secrecy rate is given by

$$R_{\mathrm{AF}}(\mathbf{F}_s, \mathbf{F}_r) = \frac{1}{2}\left[I(\mathbf{x}_s; \mathbf{y}_d) - I(\mathbf{x}_s; \mathbf{y}_e)\right]^+, \tag{4.45}$$

where

$$I(\mathbf{x}_s; \mathbf{y}_d) = \log\det\left(\mathbf{I}_k + \mathbf{F}_s^H\tilde{\mathbf{H}}_{sd}^H\mathbf{K}_{\tilde{\mathbf{w}}_d}^{-1}\tilde{\mathbf{H}}_{sd}\mathbf{F}_s\right)$$

$$
= \log \det \left(\mathbf{I}_k + \mathbf{F}_s^H \begin{bmatrix} \mathbf{H}_{sd} \\ \mathbf{H}_{rd}\mathbf{F}_r\mathbf{H}_{sr} \end{bmatrix}^H \begin{bmatrix} \mathbf{I}_{n_d} & \mathbf{0} \\ \mathbf{0} & \mathbf{I}_{n_d}+\mathbf{H}_{rd}\mathbf{F}_r\mathbf{F}_r^H\mathbf{H}_{rd}^H \end{bmatrix}^{-1} \begin{bmatrix} \mathbf{H}_{sd} \\ \mathbf{H}_{rd}\mathbf{F}_r\mathbf{H}_{sr} \end{bmatrix} \mathbf{F}_s \right)
$$

$$
= \log \det \Big[\mathbf{I}_k + \mathbf{F}_s^H \mathbf{H}_{sd}^H \mathbf{H}_{sd} \mathbf{F}_s
$$
$$
+ \mathbf{F}_s^H \mathbf{H}_{sr}^H \mathbf{F}_r^H \mathbf{H}_{rd}^H (\mathbf{I} + \mathbf{H}_{rd}\mathbf{F}_r\mathbf{F}_r^H\mathbf{H}_{rd}^H)^{-1} \mathbf{H}_{rd}\mathbf{F}_r\mathbf{H}_{sr}\mathbf{F}_s \Big]
$$

and, similarly,

$$
I(\mathbf{x}_s; \mathbf{y}_e) = \log \det \Big[\mathbf{I}_k + \mathbf{F}_s^H \mathbf{H}_{se}^H \mathbf{H}_{se} \mathbf{F}_s
$$
$$
+ \mathbf{F}_s^H \mathbf{H}_{sr}^H \mathbf{F}_r^H \mathbf{H}_{re}^H (\mathbf{I} + \mathbf{H}_{re}\mathbf{F}_r\mathbf{F}_r^H\mathbf{H}_{re}^H)^{-1} \mathbf{H}_{re}\mathbf{F}_r\mathbf{H}_{sr}\mathbf{F}_s \Big].
$$

The task is to determine the source precoder $\mathbf{F}_s$ and the relay precoder $\mathbf{F}_r$ to maximize the achievable secrecy rate subject to the individual power constraints $\operatorname{tr}(\mathbf{F}_s\mathbf{F}_s^H) = P_s$ and $\operatorname{tr}(\mathbf{F}_r(\mathbf{H}_{sr}\mathbf{F}_s\mathbf{F}_s^H\mathbf{H}_{sr}^H + \mathbf{I}_{n_r})\mathbf{F}_r^H) = P_r$ (and possibly a total power constraint $P_s + P_r = \bar{P}$). Notice that $\mathbf{F}_s$ and $\mathbf{F}_r$ are coupled not only through the achievable secrecy rate expression (which is non-convex with respect to $\mathbf{F}_s$ and $\mathbf{F}_r$) but also through the power constraint at the relay. Therefore, the joint optimization of $\mathbf{F}_s$ and $\mathbf{F}_r$ is in general difficult to solve, even for conventional non-secrecy applications without eavesdroppers. In conventional non-secrecy applications, many works consider the alternative optimization approach where $\mathbf{F}_s$ and $\mathbf{F}_r$ are optimized alternatively while keeping the other fixed. However, in secrecy applications, this problem is difficult even for the optimization of $\mathbf{F}_s$ or $\mathbf{F}_r$ individually and, thus, one must often resort to suboptimal solutions using, e.g., the zero-forcing approach. In the following, we show as an example a simpler and more tractable scenario to illustrate the relay advantages.

Let us consider a special case where the source, the destination, and the eavesdropper are all equipped with only a single antenna (i.e., $n_s = n_d = n_e = 1$). The n_r relay antennas are assumed to be distributed over n_r relays, each with only a single antenna. This is similar to the problem considered in the DF scenario and was previously examined in [3]. In this scenario, the channel matrices $\mathbf{H}_{sr}$, $\mathbf{H}_{rd}$, and $\mathbf{H}_{re}$ in (4.1) and (4.2) (or in (4.39) and (4.41)) are replaced with $\mathbf{h}_{sr} \in \mathbb{C}^{n_r \times 1}$, $\mathbf{h}_{rd} \in \mathbb{C}^{1 \times n_r}$, and $\mathbf{h}_{re} \in \mathbb{C}^{1 \times n_r}$, respectively, and with $\mathbf{H}_{sd}$ and $\mathbf{H}_{se}$ replaced with scalars h_{sd} and h_{se}, respectively. In particular, the ith entry in the vectors $\mathbf{h}_{sr} = [h_{s,r_1}, \ldots, h_{s,r_{n_r}}]^T$, $\mathbf{h}_{rd} = [h_{r_1,d}, \ldots, h_{r_{n_r},d}]$, and $\mathbf{h}_{re} = [h_{r_1,e}, \ldots, h_{r_{n_r},e}]$, namely, h_{s,r_i}, $h_{r_i,d}$, and $h_{r_i,e}$, represent the channels from the source to relay i, from relay i to the destination, and from relay i to the eavesdropper, respectively.

In the problem under consideration, each relay is only able to observe and process a single entry in the received signal vector $\mathbf{y}_r^{(1)} = [y_{r_1}^{(1)}, \ldots, y_{r_{n_r}}^{(1)}]^T$, e.g., $y_{r_i}^{(1)}$ for relay i. Hence, the relay precoding matrix takes on the form of a diagonal matrix

$$\mathbf{F}_r = \text{diag}(\mathbf{f}_r) = \text{diag}(f_{r_1}, \ldots, f_{r_{n_r}}), \tag{4.46}$$

where $\mathbf{f}_r \triangleq [f_{r_1}, \ldots, f_{r_{n_r}}]^T$ with f_{r_i} being the gain imposed by relay i. In this case, the relay transmit power can be written as

$$P_r = \text{tr}(P_s \mathbf{F}_r (\mathbf{h}_{sr} \mathbf{h}_{sr}^H + \mathbf{I}_{n_r}) \mathbf{F}_r^H) = \mathbf{f}_r^H \mathbf{D}_{sr} \mathbf{f}_r \tag{4.47}$$

where $\mathbf{D}_{sr} \triangleq P_s \text{diag}(\mathbf{h}_{sr})^H \text{diag}(\mathbf{h}_{sr}) + \mathbf{I}_{n_r}$. Following (4.45), the achievable secrecy rate can be written as

$$\begin{aligned}
R_{\text{AF}}(P_s, \mathbf{F}_r) &= \frac{1}{2} \left[\log \frac{1 + P_s |h_{sd}|^2 + \frac{P_s \mathbf{h}_{sr}^H \mathbf{F}_r^H \mathbf{h}_{rd}^H \mathbf{h}_{rd} \mathbf{F}_r \mathbf{h}_{sr}}{1 + \mathbf{h}_{rd} \mathbf{F}_r \mathbf{F}_r^H \mathbf{h}_{rd}}}{1 + P_s |h_{se}|^2 + \frac{P_s \mathbf{h}_{sr}^H \mathbf{F}_r^H \mathbf{h}_{re}^H \mathbf{h}_{re} \mathbf{F}_r \mathbf{h}_{sr}}{1 + \mathbf{h}_{re} \mathbf{F}_r \mathbf{F}_r^H \mathbf{h}_{re}}} \right]^+ \\
&= \frac{1}{2} \left[\log \frac{1 + P_s |h_{sd}|^2 + \frac{\mathbf{f}_r^H \mathbf{R}_{srd} \mathbf{f}_r}{1 + \mathbf{f}_r^H \mathbf{R}_{rd} \mathbf{f}_r}}{1 + P_s |h_{se}|^2 + \frac{\mathbf{f}_r^H \mathbf{R}_{sre} \mathbf{f}_r}{1 + \mathbf{f}_r^H \mathbf{R}_{re} \mathbf{f}_r}} \right]^+ \tag{4.48}
\end{aligned}$$

where $\mathbf{R}_{srd} \triangleq P_s \text{diag}(\mathbf{h}_{sr})^H \mathbf{h}_{rd}^H \mathbf{h}_{rd} \text{diag}(\mathbf{h}_{sr})$, $\mathbf{R}_{rd} \triangleq \text{diag}(\mathbf{h}_{rd})^H \text{diag}(\mathbf{h}_{rd})$, $\mathbf{R}_{sre} \triangleq P_s \text{diag}(\mathbf{h}_{sr})^H \mathbf{h}_{re}^H \mathbf{h}_{re} \text{diag}(\mathbf{h}_{sr})$, and $\mathbf{R}_{re} \triangleq \text{diag}(\mathbf{h}_{re})^H \text{diag}(\mathbf{h}_{re})$. By adopting the relay power expression in (4.47), the achievable rate can be further expressed as

$$R_{\text{AF}}(P_s, P_r, \mathbf{f}_r) = \frac{1}{2} \left[\log \left(\frac{\mathbf{f}_r^H \tilde{\mathbf{R}}_{srd} \mathbf{f}_r}{\mathbf{f}_r^H \tilde{\mathbf{R}}_{sre} \mathbf{f}_r} \cdot \frac{\mathbf{f}_r^H \tilde{\mathbf{R}}_{re} \mathbf{f}_r}{\mathbf{f}_r^H \tilde{\mathbf{R}}_{rd} \mathbf{f}_r} \right) \right]^+ \tag{4.49}$$

where $\tilde{\mathbf{R}}_{rd} \triangleq P_r^{-1} \mathbf{D}_{sr} + \mathbf{R}_{rd}$, $\tilde{\mathbf{R}}_{re} \triangleq P_r^{-1} \mathbf{D}_{sr} + \mathbf{R}_{re}$, $\tilde{\mathbf{R}}_{srd} \triangleq (1 + P_s |h_{sd}|^2) \tilde{\mathbf{R}}_{rd} + \mathbf{R}_{srd}$, and $\tilde{\mathbf{R}}_{sre} \triangleq (1 + P_s |h_{se}|^2) \tilde{\mathbf{R}}_{re} + \mathbf{R}_{sre}$. The optimization problem can then be formulated as follows [3]:

$$\max_{P_s, \mathbf{f}_r} \frac{\mathbf{f}_r^H \tilde{\mathbf{R}}_{srd} \mathbf{f}_r}{\mathbf{f}_r^H \tilde{\mathbf{R}}_{sre} \mathbf{f}_r} \cdot \frac{\mathbf{f}_r^H \tilde{\mathbf{R}}_{re} \mathbf{f}_r}{\mathbf{f}_r^H \tilde{\mathbf{R}}_{rd} \mathbf{f}_r} \tag{4.50a}$$

$$\text{subject to } P_s + P_r = \bar{P}, \tag{4.50b}$$

$$\mathbf{f}_r^H \mathbf{D}_{sr} \mathbf{f}_r = P_r. \tag{4.50c}$$

Notice that the objective is the product of two generalized Rayleigh quotients [5]. The beamforming directions that maximize the first and second terms, respectively, are given by $\tilde{\mathbf{f}}_r^{(1)} = \psi_{\max}(\tilde{\mathbf{R}}_{srd}, \tilde{\mathbf{R}}_{sre})$ and $\tilde{\mathbf{f}}_r^{(2)} = \psi_{\max}(\tilde{\mathbf{R}}_{re}, \tilde{\mathbf{R}}_{rd})$, where $\psi_{\max}(\mathbf{A}, \mathbf{B})$ denotes the generalized eigenvector corresponding to the maximum generalized eigenvalue of the matrix pair $(\mathbf{A}, \mathbf{B})$. By taking into consideration the relay power constraint P_r, the relay beamforming vectors are then given by respectively

$$\mathbf{f}_r^{(1)} = \sqrt{\frac{P_r}{(\tilde{\mathbf{f}}_r^{(1)})^H \mathbf{D}_{sr} \tilde{\mathbf{f}}_r^{(1)}}} \tilde{\mathbf{f}}_r^{(1)} \quad \text{and} \quad \mathbf{f}_r^{(2)} = \sqrt{\frac{P_r}{(\tilde{\mathbf{f}}_r^{(2)})^H \mathbf{D}_{sr} \tilde{\mathbf{f}}_r^{(2)}}} \tilde{\mathbf{f}}_r^{(2)}.$$

However, finding the beamforming direction that maximizes the product of the two terms is difficult. In [3], the authors proposed to optimize the first term only, namely, to choose $\mathbf{f}_r^* = \mathbf{f}_r^{(1)}$. This solution is shown to closely approximate the optimal solution when the channel gains between the relays and the destination are approximately the same as those between the relays and the eavesdropper, or when the signal power at the relay is much larger than that at the destination. The optimal power allocation between P_s and P_r can then be found by a simple line search of $P_s \in [0, \bar{P}]$ or by using the conventional random search algorithm [3, 6].

4.2 Distributed Secrecy Beamforming and Precoding with Untrusted Relays

One of the main concerns of using relay-assisted transmissions in secrecy applications is the trustworthiness of the relay or relays. This was first examined from an information-theoretic perspective in [7, 2], where an untrusted relay is treated as an adversary that assists the source transmission but also attempts to intercept the confidential message during the process. By treating the relay as the eavesdropper, the achievable secrecy rate can also be derived under the constraint that the equivocation rate at the relay approaches the entropy of the messages [7, 2]. Notice that the DF relaying strategy is not applicable in this case since the untrusted relay is not allowed to decode the messages. Therefore, AF and compress-and-forward (CF) relaying strategies have been considered the most in literature for this scenario [7, 2]. Intuitively, advantages can still be gained by employing untrusted relays when the combined signal from the direct and relay paths yield a better effective signal quality at the destination than that at the relay (i.e., eavesdropper). In the following, we consider specifically the case of AF relaying and introduce approaches to determine efficiently the source and relay precoders.

Let us consider a multi-antenna wireless relay system with an untrusted relay, as depicted in Fig. 4.2. The source, the relay, and the destination are equipped with n_s, n_r, and n_d antennas, respectively. Here, the relay helps to forward the source signal using an AF strategy but at the same time serves as the sole eavesdropper in the system. A two-phase transmission is also adopted here, similar to that in Sect. 4.1, but without a separate eavesdropper. In Phase 1, the source transmits the signal $\mathbf{x}_s = \mathbf{F}_s \mathbf{u}$, where $\mathbf{F}_s \in \mathbb{C}^{n_s \times k}$ is the source precoder, $\mathbf{u} \in \mathcal{CN}(0, \mathbf{I}_k)$ is the encoded symbol vector, and k is the number of signal dimensions. The transmit power at the source is given by $P_s = \mathbb{E}[\|\mathbf{x}_s\|^2] = \mathrm{tr}(\mathbf{F}_s \mathbf{F}_s^H)$. The received signals at the relay (i.e., eavesdropper) and the destination are given by

$$\mathbf{y}_r^{(1)} = \mathbf{H}_{sr}\mathbf{x}_s + \mathbf{w}_r^{(1)}, \tag{4.51a}$$

$$\mathbf{y}_d^{(1)} = \mathbf{H}_{sd}\mathbf{x}_s + \mathbf{w}_d^{(1)}, \tag{4.51b}$$

where the notations follow that of (4.1). In Phase 2, the relay forwards a linearly amplified version of $\mathbf{y}_r^{(1)}$ to the destination. The signal transmitted by the relay is written as $\mathbf{x}_r = \mathbf{F}_r\mathbf{y}_r^{(1)}$, where $\mathbf{F}_r \in \mathbb{C}^{n_r \times n_r}$ is the relay precoder. The transmit power at the relay is given by $P_r \triangleq \mathrm{E}[\|\mathbf{x}_r\|^2] = \mathrm{tr}(\mathbf{F}_r(\mathbf{H}_{sr}\mathbf{F}_s\mathbf{F}_s^H\mathbf{H}_{sr}^H + \mathbf{I}_{n_r})\mathbf{F}_r^H)$. The corresponding received signal at the destination is given by

$$\mathbf{y}_d^{(2)} = \mathbf{H}_{rd}\mathbf{x}_r + \mathbf{w}_d^{(2)}, \tag{4.52}$$

where the notations follow that of (4.2a). The signals received in the two phases at the destination can be combined into the effective received signal vector in (4.39). Since the relay is treated as the eavesdropper, the achievable secrecy rate is given by

$$R_{\mathrm{AF,u}}(\mathbf{F}_s, \mathbf{F}_r) = \frac{1}{2}\left[I(\mathbf{x}_s; \mathbf{y}_d) - I(\mathbf{x}_s; \mathbf{y}_r^{(1)})\right], \tag{4.53}$$

where

$$I(\mathbf{x}_s; \mathbf{y}_d) = \log\det\left[\mathbf{I}_k + \mathbf{F}_s^H\mathbf{H}_{sd}^H\mathbf{H}_{sd}\mathbf{F}_s + \mathbf{F}_s^H\mathbf{H}_{sr}^H\mathbf{F}_r^H\mathbf{H}_{rd}^H \\ \cdot (\mathbf{I} + \mathbf{H}_{rd}\mathbf{F}_r\mathbf{F}_r^H\mathbf{H}_{rd}^H)^{-1}\mathbf{H}_{rd}\mathbf{F}_r\mathbf{H}_{sr}\mathbf{F}_s\right] \tag{4.54}$$

and

$$I(\mathbf{x}_s; \mathbf{y}_r^{(1)}) = \log\det(\mathbf{I}_{n_r} + \mathbf{H}_{sr}\mathbf{F}_s\mathbf{F}_s^H\mathbf{H}_{sr}^H). \tag{4.55}$$

The source and relay precoders can then be found to maximize the achievable secrecy rate. However, the joint optimization is again difficult due to similar reasons as before. For tractability, most works adopt the hill-climbing or alternating optimization approaches, where $\mathbf{F}_s$ and $\mathbf{F}_r$ are optimized in turn while keeping the other fixed. Notice that, when $\mathbf{F}_r$ is fixed, the search for the optimal source precoder $\mathbf{F}_s$ becomes

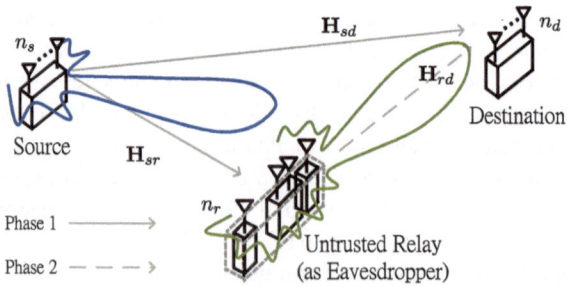

Fig. 4.2 Illustration of distributed secrecy beamforming with an untrusted relay

identical to that considered in Sect. 3.1.2 with $\mathbf{H}_{sd}$ and $\mathbf{H}_{re}$ being the channel matrices to the destination and the eavesdropper, respectively. However, since in general multiple signal dimensions can be used, explicit solutions of $\mathbf{F}_s$ can only be found under the power covariance constraint and must resort to numerical methods under the total power constraint.

Let us consider the special case, previously studied in [8], where the number of signal dimensions k is equal to 1. That is, the source is restricted to using a rank-1 beamforming vector, denoted by $\mathbf{f}_s$. In this case, the signal transmitted by the source can be written as $\mathbf{x}_s = \mathbf{f}_s u$, where $u \sim \mathcal{CN}(0, 1)$ is the encoded symbol. The achievable secrecy rate can then be written as

$$R_{\text{AF,u}}(\mathbf{F}_s, \mathbf{F}_r) = \frac{1}{2} \log \frac{1 + \mathbf{f}_s^H \mathbf{Q}_{sd}(\mathbf{F}_r)\mathbf{f}_s}{1 + \mathbf{f}_s^H \mathbf{Q}_{sr}\mathbf{f}_s}, \tag{4.56}$$

where $\mathbf{Q}_{sd}(\mathbf{F}_r) \triangleq \mathbf{H}_{sd}^H \mathbf{H}_{sd} + \mathbf{H}_{sr}^H \mathbf{F}_r^H \mathbf{H}_{rd}^H (\mathbf{I} + \mathbf{H}_{rd}\mathbf{F}_r\mathbf{F}_r^H \mathbf{H}_{rd}^H)^{-1}\mathbf{H}_{rd}\mathbf{F}_r\mathbf{H}_{sr}$ and $\mathbf{Q}_{sr} \triangleq \mathbf{H}_{sr}^H \mathbf{H}_{sr}$. Notice that only the numerator inside the log function depends on the relay precoder $\mathbf{F}_r$. Following the method proposed in [8], we consider the alternating optimization approach where the source beamformer $\mathbf{f}_s$ is first found while $\mathbf{F}_r$ is fixed and vice versa. Notice that, within each optimization, the power constraints $\mathbf{f}_s^H \mathbf{f}_s \leq P_s$ and $\text{tr}(\mathbf{F}_r(\mathbf{H}_{sr}\mathbf{f}_s\mathbf{f}_s^H \mathbf{H}_{sr}^H + \mathbf{I}_{n_r})\mathbf{F}_r^H) \leq P_r$ must be satisfied.

Let us first consider the optimization of $\mathbf{f}_s$ for given $\mathbf{F}_r$. Notice that, when $\mathbf{F}_r$ is fixed, the relay power constraint can be written explicitly as a constraint on $\mathbf{f}_s$, i.e., $\mathbf{f}_s^H \mathbf{H}_{sr}^H \mathbf{F}_r^H \mathbf{F}_r\mathbf{H}_{sr}\mathbf{f}_s \leq P_r - \text{tr}(\mathbf{F}_r^H \mathbf{F}_r)$. Hence, the optimization problem for $\mathbf{f}_s$ can be formulated as the following fractional quadratically constrained quadratic programming (QCQP) problem:

$$\max_{\mathbf{f}_s} \quad \frac{1 + \mathbf{f}_s^H \mathbf{Q}_{sd}(\mathbf{F}_r)\mathbf{f}_s}{1 + \mathbf{f}_s^H \mathbf{Q}_{sr}\mathbf{f}_s} \tag{4.57a}$$

$$\text{subject to } \mathbf{f}_s^H \mathbf{f}_s \leq P_s, \tag{4.57b}$$

$$\mathbf{f}_s^H \mathbf{H}_{sr}^H \mathbf{F}_r^H \mathbf{F}_r\mathbf{H}_{sr}\mathbf{f}_s \leq P_r - \text{tr}(\mathbf{F}_r^H \mathbf{F}_r). \tag{4.57c}$$

Since the relay precoder $\mathbf{F}_r$ is fixed, the relay power constraint can be written explicitly as a constraint on $\mathbf{f}_s$, i.e., $\mathbf{f}_s^H \mathbf{H}_{sr}^H \mathbf{F}_r^H \mathbf{F}_r\mathbf{H}_{sr}\mathbf{f}_s \leq P_r - \text{tr}(\mathbf{F}_r^H \mathbf{F}_r)$. The optimal solution to the problem in (4.57) can be found using an algorithm previously proposed in [8].

Specifically, using the approach in [9], the problem in (4.57) is first relaxed into a fractional semi-definite programming problem, which is then transformed into a semi-definite programming problem (SDP) using the Charnes-Cooper transformation technique. The SDP is given as follows [8]:

$$\max_{\mathbf{Z},\nu} \ \mathrm{tr}(\tilde{\mathbf{Q}}_{sd}\mathbf{Z}) \tag{4.58a}$$

$$\text{subject to} \ \ \mathrm{tr}(\tilde{\mathbf{Q}}_{sr}\mathbf{Z}) = 1, \tag{4.58b}$$

$$\mathrm{tr}(\mathbf{A}_1\mathbf{Z}) \le \nu P_s \tag{4.58c}$$

$$\mathrm{tr}(\mathbf{A}_2\mathbf{Z}) \le \nu[P_r - \mathrm{tr}(\mathbf{F}_r^H \mathbf{F}_r)] \tag{4.58d}$$

$$\mathrm{tr}(\mathbf{A}_3\mathbf{Z}) = \nu \tag{4.58e}$$

$$\nu \ge 0. \tag{4.58f}$$

where $\tilde{\mathbf{Q}}_{sd} \triangleq \mathrm{diag}(\mathbf{Q}_{sd}, 1)$, $\tilde{\mathbf{Q}}_{sr} \triangleq \mathrm{diag}(\mathbf{Q}_{sr}, 1)$, $\mathbf{A}_1 \triangleq \mathrm{diag}(\mathbf{I}_{n_s}, 0)$, $\mathbf{A}_2 \triangleq$ $\mathrm{diag}(\mathbf{H}_{sr}^H\mathbf{F}_r^H\mathbf{F}_r\mathbf{H}_{sr}, 0)$ and $\mathbf{A}_3 \triangleq \mathrm{diag}(\mathbf{O}_{n_s}, 1)$. Here, $\mathbf{O}_{n_s} \in \mathbb{C}^{n_s \times n_s}$ is an all zero matrix. The following algorithm can then be used to obtain an optimal solution for $\mathbf{f}_s$.

Algorithm for Finding Optimal $\mathbf{f}_s$ [8]

1. Find the optimal solution $(\mathbf{Z}^*, \nu^*)$ to (4.58) and the maximum objective value η^*. Let $\mathbf{X}^* = \mathbf{Z}^*/\nu^*$.
2. If $\mathrm{rank}(\mathbf{X}^*) = 1$, then decompose $\mathbf{X}^*$ as $\mathbf{X}^* = \mathbf{x}^*(\mathbf{x}^*)^H$ and go to step 4.
3. If $\mathrm{rank}(\mathbf{X}^*) \ge 2$, find $\mathbf{X}^*$ such that

$$(\mathbf{x}^*)^H (\tilde{\mathbf{Q}}_{sd} - \eta^*\tilde{\mathbf{Q}}_{sr})\mathbf{x}^* = \mathrm{tr}((\tilde{\mathbf{Q}}_{sd} - \eta^*\tilde{\mathbf{Q}}_{sr})\mathbf{X}^*)$$

$$(\mathbf{x}^*)^H \mathbf{A}_i\mathbf{x}^* = \mathrm{tr}(\mathbf{A}_i\mathbf{X}^*), \quad \text{for } i = 1, 2, 3.$$

4. Let $\mathbf{x}^* = [\mathbf{v}^T, t]^T$ and determine the optimal solution as $\mathbf{f}_s^* = \mathbf{v}/t$.

Next, let us consider the optimization of $\mathbf{F}_r$ for given $\mathbf{f}_s$. Notice that, when $\mathbf{f}_s$ is fixed, only the numerator in (4.56), i.e., the mutual information at the destination, depends on $\mathbf{F}_r$. The problem becomes identical to the conventional relay precoding problem and is formulated as follows:

$$\max_{\mathbf{F}_r} \ \ \log(1 + \mathbf{f}_s^H \mathbf{Q}_{sd}(\mathbf{F}_r)\mathbf{f}_s) \tag{4.59a}$$

$$\text{subject to} \ \ \mathrm{tr}(\mathbf{F}_r(\mathbf{H}_{sr}\mathbf{f}_s\mathbf{f}_s^H \mathbf{H}_{sr}^H + \mathbf{I}_{n_r})\mathbf{F}_r^H) \le P_r. \tag{4.59b}$$

This problem can be solved following the derivations in [8, 10] and is summarized as follows.

Let $\mathbf{H}_{rd} = \mathbf{U}_{rd}\mathbf{\Sigma}_{rd}\mathbf{V}_{rd}^H$ be the singular value decomposition (SVD) of $\mathbf{H}_{rd}$, where $\mathbf{U}_{rd} \in \mathbb{C}^{n_d \times p}$ and $\mathbf{V}_{rd} \in \mathbb{C}^{n_r \times p}$ are semi-unitary matrices, $\mathbf{\Sigma}_{rd} \in \mathbb{C}^{p \times p}$ is a diagonal matrix of singular values of $\mathbf{H}_{rd}$, and $p = \mathrm{rank}(\mathbf{H}_{rd})$. The matrices are chosen such that the singular values on the diagonal of $\mathbf{\Sigma}_{rd}$ are in descending order. It has been shown in [10] that the optimal relay precoder should take on the following structure

$$\mathbf{F}_r = \mathbf{V}_{rd}\mathbf{a}\mathbf{u}_{sr}^H \tag{4.60}$$

where $\mathbf{u}_{sr} \triangleq \mathbf{H}_{sr}\mathbf{f}_s/\|\mathbf{H}_{sr}\mathbf{f}_s\|$ is the effective source-to-relay channel given the source beamformer $\mathbf{f}_s$ and $\mathbf{a} \in \mathbb{C}^{p \times 1}$ is a vector chosen to satisfy the relay power constraint.

Moreover, by the matrix inversion lemma, the second term inside the logarithmic function in (4.59) can be expressed as

$$
\begin{aligned}
&\mathbf{f}_s^H \mathbf{Q}_{sd}(\mathbf{F}_r) \mathbf{f}_s \\
&= \mathbf{f}_s^H \mathbf{H}_{sd}^H \mathbf{H}_{sd} \mathbf{f}_s + \mathbf{f}_s^H \mathbf{H}_{sr}^H \mathbf{F}_r^H \mathbf{H}_{rd}^H (\mathbf{I} + \mathbf{H}_{rd} \mathbf{F}_r \mathbf{F}_r^H \mathbf{H}_{rd}^H)^{-1} \mathbf{H}_{rd} \mathbf{F}_r \mathbf{H}_{sr} \mathbf{f}_s \\
&= \mathbf{f}_s^H (\mathbf{H}_{sd}^H \mathbf{H}_{sd} + \mathbf{H}_{sr}^H \mathbf{H}_{sr}) \mathbf{f}_s - \mathbf{f}_s^H \mathbf{H}_{sr}^H (\mathbf{I} + \mathbf{F}_r^H \mathbf{H}_{rd}^H \mathbf{H}_{rd} \mathbf{F}_r)^{-1} \mathbf{H}_{sr} \mathbf{f}_s \\
&= \mathbf{f}_s^H (\mathbf{H}_{sd}^H \mathbf{H}_{sd} + \mathbf{H}_{sr}^H \mathbf{H}_{sr}) \mathbf{f}_s - \frac{\|\mathbf{H}_{sr} \mathbf{f}_s\|^2}{1 + \mathbf{a}^H \mathbf{\Sigma}_{rd}^2 \mathbf{a}},
\end{aligned}
$$

where the last equality is obtained by replacing $\mathbf{F}_r$ with its optimal structure in (4.60). By (4.60), the relay power constraint can also be written as

$$
\mathbf{a}^H \mathbf{a}(1 + \|\mathbf{H}_{sr} \mathbf{f}_s\|^2) \le P_r. \tag{4.61}
$$

Since, only the denominator in the third term depends on $\mathbf{F}_r$, the optimization problem can be reduced to

$$
\max_{\mathbf{a}} \ \mathbf{a}^H \mathbf{\Sigma}_{rd}^H \mathbf{\Sigma}_{rd} \mathbf{a} \tag{4.62a}
$$

$$
\text{subject to } \|\mathbf{a}\|^2 \le P_r / (1 + \|\mathbf{H}_{sr} \mathbf{f}_s\|^2). \tag{4.62b}
$$

Since the singular values of $\mathbf{\Sigma}_{rd}$ are listed in descending order, the solution to the optimization problem in (4.62) is given by

$$
\mathbf{a}^* = \sqrt{\frac{P_r}{1 + \|\mathbf{H}_{sr} \mathbf{f}_s\|^2}} \mathbf{e}_1, \tag{4.63}
$$

where $\mathbf{e}_1$ is a $p \times 1$ vector with 1 in the first entry and 0 in the other entries. The optimal relay precoder is thus given by

$$
\mathbf{F}_r = \frac{\sqrt{P_r / (1 + \|\mathbf{H}_{sr} \mathbf{f}_s\|^2)}}{\|\mathbf{H}_{sr} \mathbf{f}_s\|} \mathbf{v}_{rd,1} \mathbf{f}_s^H \mathbf{H}_{sr}^H, \tag{4.64}
$$

where $\mathbf{v}_{rd,1}$ represents the first column of $\mathbf{V}_{rd}$. This solution indicates that the optimal relay beamformer should be matched to both the effective source-to-relay channel $\mathbf{u}_{sr}$ and the largest singular mode of $\mathbf{H}_{rd}$.

The two optimization problems mentioned above, namely, (i) the optimization of $\mathbf{f}_s$ given $\mathbf{F}_r$ and (ii) the optimization of $\mathbf{F}_r$ given $\mathbf{f}_s$, can then be performed iteratively until no further improvement can be obtained in the achievable secrecy rate. This approach does not necessarily yield the globally optimal solution but provides a tractable way to obtain effective designs of $\mathbf{f}_s$ and $\mathbf{F}_r$.

Alternatively, one can also consider a special case, similar to that in Sect. 4.1.2, where the source and the destination are both equipped with only a single antenna (i.e., $n_s = n_d = 1$). The n_r relay antennas are assumed to be distributed over n_r single-antenna relays. These relays are untrusted and, thus, are treated as the eavesdroppers in this case. Here, the channel matrices $\mathbf{H}_{sr}$ and $\mathbf{H}_{rd}$ are replaced with $\mathbf{h}_{sr} \in \mathbb{C}^{n_r \times 1}$ and $\mathbf{h}_{rd} \in \mathbb{C}^{1 \times n_r}$, respectively, and $\mathbf{H}_{sd}$ is replaced with the scalar h_{sd}. Again, the ith entry in the vectors $\mathbf{h}_{sr}$ and $\mathbf{h}_{rd}$, which are denoted by h_{s,r_i} and $h_{r_i,d}$, represent the channels from the source to relay i and from relay i to the destination, respectively. The signal transmitted by the source is represented by a scalar $x_s = \sqrt{P_s}u$, where $u \in \mathcal{CN}(0, 1)$. Therefore, in Phase 1, the signal received at the relay (i.e., eavesdropper) and the destination can be written as

$$\mathbf{y}_r^{(1)} = \mathbf{h}_{sr}x_s + \mathbf{w}_r^{(1)}, \tag{4.65a}$$

$$y_d^{(1)} = h_{sd}x_s + w_d^{(1)}. \tag{4.65b}$$

In Phase 2, each relay forwards a linearly amplified version of its locally received signal to the destination. Since relay i is only able to receive the ith entry in $\mathbf{y}_r^{(1)}$, the signal transmitted by the relays can then be written as $\mathbf{x}_r = \mathbf{F}_r\mathbf{y}_r^{(1)}$, where $\mathbf{F}_r = \mathrm{diag}(\mathbf{f}_r) = \mathrm{diag}(f_{r_1}, \ldots, f_{r_{n_r}}) \in \mathbb{C}^{n_r \times n_r}$ is in the form of a diagonal matrix and $\mathbf{f}_r \triangleq [f_{r_1}, \ldots, f_{r_{n_r}}]^T$. Similar to (4.47), the transmit power at the relay is given by

$$P_r = \mathrm{tr}(\mathbf{F}_r(P_s\mathbf{h}_{sr}\mathbf{h}_{sr}^H + \mathbf{I}_{n_r})\mathbf{F}_r^H) = \mathbf{f}_r^H\mathbf{D}_r\mathbf{f}_r, \tag{4.66}$$

where $\mathbf{D}_r \triangleq P_s\mathrm{diag}(\mathbf{h}_{sr})^H\mathrm{diag}(\mathbf{h}_{sr}) + \mathbf{I}_{n_r}$. The corresponding received signal at the destination is given by

$$y_d^{(2)} = \mathbf{h}_{rd}\mathbf{x}_r + w_d^{(2)}. \tag{4.67}$$

The signals received in the two phases at the destination can be combined into the effective received signal vector

$$\mathbf{y}_d = \begin{bmatrix} y_d^{(1)} \\ y_d^{(2)} \end{bmatrix} = \begin{bmatrix} h_{sd} \\ \mathbf{h}_{rd}\mathbf{F}_r\mathbf{h}_{sr} \end{bmatrix} x_s + \begin{bmatrix} w_d^{(1)} \\ \mathbf{h}_{rd}\mathbf{F}_r\mathbf{w}_r^{(1)} + w_d^{(2)} \end{bmatrix}. \tag{4.68}$$

By assuming that the relays are non-colluding, i.e., they are not allowed to share information among each other, an achievable secrecy rate is given by

$$R_{\mathrm{AF},u}(P_s, \mathbf{F}_r) = \frac{1}{2}\left[I(x_s; \mathbf{y}_d) - \min_{i=1,\ldots,n_r} I(x_s; y_{r_i}^{(1)}) \right] \tag{4.69}$$

$$= \frac{1}{2}\log\frac{1 + P_s|h_{sd}|^2 + \frac{P_s\mathbf{h}_{sr}^H\mathbf{F}_r^H\mathbf{h}_{rd}^H\mathbf{h}_{rd}\mathbf{F}_r\mathbf{h}_{sr}}{1+\mathbf{h}_{rd}\mathbf{F}_r\mathbf{F}_r^H\mathbf{h}_{rd}}}{1 + \min\limits_{i=1,\ldots,n_r} P_s|h_{s,r_i}|^2}$$

$$= \frac{1}{2}\log\frac{1 + P_s|h_{sd}|^2 + \frac{\mathbf{f}_r^H\mathbf{R}_{srd}\mathbf{f}_r}{1+\mathbf{f}_r^H\mathbf{R}_{rd}\mathbf{f}_r}}{1 + \min\limits_{i=1,\ldots,n_r} P_s|h_{s,r_i}|^2} \tag{4.70}$$

where $\mathbf{R}_{srd} \triangleq P_s \, \text{diag}(\mathbf{h}_{sr})^H \mathbf{h}_{rd}^H \mathbf{h}_{rd} \, \text{diag}(\mathbf{h}_{sr})$ and $\mathbf{R}_{rd} \triangleq \text{diag}(\mathbf{h}_{rd})^H \text{diag}(\mathbf{h}_{rd})$. In this case, it is sufficient to find $\mathbf{f}_r$ that maximizes the ratio

$$\frac{\mathbf{f}_r^H \mathbf{R}_{srd} \mathbf{f}_r}{1 + \mathbf{f}_r^H \mathbf{R}_{rd} \mathbf{f}_r} = \frac{\mathbf{f}_r^H \mathbf{R}_{srd} \mathbf{f}_r}{\mathbf{f}_r^H \tilde{\mathbf{R}}_{rd} \mathbf{f}_r} \tag{4.71}$$

where $\tilde{\mathbf{R}}_{rd} \triangleq P_r^{-1}[P_s \, \text{diag}(\mathbf{h}_{sr})^H \, \text{diag}(\mathbf{h}_{sr}) + \mathbf{I}_{n_r}] + \mathbf{R}_{rd}$. Hence, the optimal relay beamformer is given by

$$\mathbf{f}_r = \sqrt{\frac{P_r}{(\tilde{\mathbf{f}}_r^*)^H \mathbf{D}_r \tilde{\mathbf{f}}_r^*}} \cdot \tilde{\mathbf{f}}_r^* \tag{4.72}$$

where $\tilde{\mathbf{f}}_r^* = \psi_{\max}(\mathbf{R}_{srd}, \tilde{\mathbf{R}}_{rd})$ is the optimal beamforming direction. The optimal power allocation between the source and the relay powers, i.e., P_s and P_r, respectively, can be found under the power constraint $P_s + P_r = \bar{P}$ via a simple line search of P_s over the interval $[0, \bar{P}]$ and by setting $P_r = \bar{P} - P_s$. The conventional random search algorithm [3, 6] can also be adopted as mentioned in the previous section.

Example 4.1 The secrecy rate and secrecy outage performances for the case of untrusted relays are shown in Fig. 4.3 for varying relay positions. Here, we consider the relay wiretap channel in Fig. 4.1, where $n_s = n_d = n_e = 1$ and $n_r = 2$. All channel coefficients are assumed to be i.i.d. $\mathcal{CN}(0, 1)$ and the path loss exponent is $\alpha = 2$. Here, the relay is treated as the eavesdropper and the positions of the source and destination are set as $(0, 0)$ and $(1, 0)$, respectively. Two cases are considered: the case without relay assistance [which reduces to the conventional single-input single-output multi-antenna eavesdropper (SISOME) scenario] and the case where the relay assists in forwarding the source's message using the AF strategy even though it is untrusted. Notice that the first scheme requires only a one-phase transmission whereas the latter scheme requires a two-phase transmission. The relay position is varied from $(0, 0.02)$ to $(2, 0.02)$ and the transmit SNR is fixed as 10 dB. The total power constraint among source and relays is considered. In Fig. 4.3a, one can observe that AF transmissions by untrusted relays can provide better performance than direct transmission when the relay is close to the destination. In Fig. 4.3b, one can observe that secrecy outage is minimized by placing the relay at position $(0.7, 0.02)$. Here, the destination benefits the most from both the source and the relay without revealing too much information to the relay (i.e., the eavesdropper).

4.3 Distributed Secrecy Beamforming and Precoding with Artificial Noise

Relay beamforming or precoding can help increase the achievable secrecy rate by enhancing the received signal quality at the destination or by reducing the signal strength at the eavesdropper. However, information forwarding by relays may not

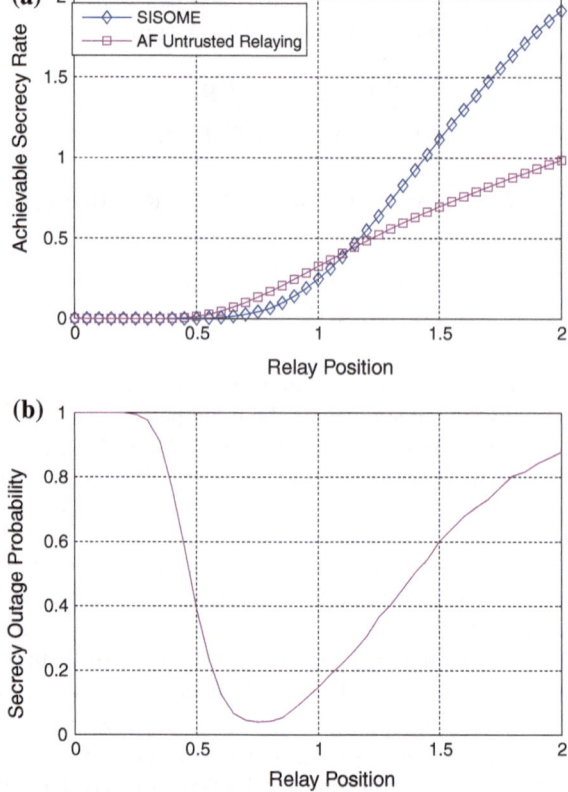

Fig. 4.3 Achievable secrecy rate and secrecy outage probability for the case with untrusted relays. Here, the number of antennas are $n_s = n_d = n_e = 1$ and $n_e = 2$. **a** Varying relay position with fixed SNR, **b** Varying relay position with fixed SNR

always be beneficial since it requires a two-phase transmission and, thus, the bandwidth efficiency is typically reduced by one half. In addition, information forwarding by relays can also be inefficient when the receive signal quality at the relays is low or when the relays are located close to the eavesdropper. In these cases, it may be better for relays to utilize part (or all) of its resources for AN emission, instead of for information forwarding, since degrading the signal quality at the eavesdropper also effectively increases the achievable secrecy rate. Also, similar to the case without relays (c.f. Chap. 3), the use of AN is especially effective when the eavesdropper's channel is not known perfectly at the transmitter.

In this section, we consider two cases: (i) the case where relays act as pure cooperative jammers and emit only AN signals to jam the eavesdropper's reception [3] and (ii) the more general case where both AN and information signals are emitted by the relays [11, 12]. When the relays act as pure cooperative jammers, only a single phase transmission is required since no information needs to be received by the relay.

When both AN and information signals are transmitted, the relay precoder and AN covariance matrix must be jointly designed to simultaneously enhance information signal and reduce AN interference at the destination. These issues are discussed in the following.

4.3.1 Relays as Pure Cooperative Jammers

In this section, we assume that the relays act as pure cooperative jammers and emit only jamming signals (instead of information signals) to disrupt the reception at the eavesdropper. In this case, the relays need not listen to the signal from the source and, thus, a single-phase transmission is sufficient.

Specifically, let us consider a multi-antenna wireless system with a source, a destination, an eavesdropper, and a relay acting as a pure cooperative jammer, as illustrated in Fig. 4.4. The terminals are equipped with n_s, n_d, n_e, and n_r antennas, respectively. Here, only the source emits the information signal whereas the relay emits only AN. The signal transmitted by the source can be written as $\mathbf{x}_s = \mathbf{F}_s \mathbf{u}$, where $\mathbf{F}_s \in \mathbb{C}^{n_s \times k}$ is the source precoder, $\mathbf{u} \in \mathcal{CN}(0, \mathbf{I}_k)$ is the encoded symbol vector, and k is the number of signal dimensions. The resulting input covariance matrix is given by $\mathbf{K}_{\mathbf{x}_s} \triangleq \mathrm{E}[\mathbf{x}_s \mathbf{x}_s^H] = \mathbf{F}_s \mathbf{F}_s^H$. The AN emitted by the relay can be represented by the Gaussian vector $\mathbf{a} \in \mathcal{CN}(0, \mathbf{K}_a)$. By considering only a single-phase transmission, the received signals at the destination and the eavesdropper can be written as

$$\mathbf{y}_d = \mathbf{H}_{sd}\mathbf{x}_s + \mathbf{H}_{rd}\mathbf{a} + \mathbf{w}_d, \tag{4.73a}$$

$$\mathbf{y}_e = \mathbf{H}_{se}\mathbf{x}_s + \mathbf{H}_{re}\mathbf{a} + \mathbf{w}_e \tag{4.73b}$$

where $\mathbf{w}_d \in \mathcal{CN}(0, \mathbf{I}_{n_d})$ and $\mathbf{w}_e \in \mathcal{CN}(0, \mathbf{I}_{n_d})$ are the AWGN at the destination and the eavesdropper, respectively. The equivalent noise covariance matrices at the

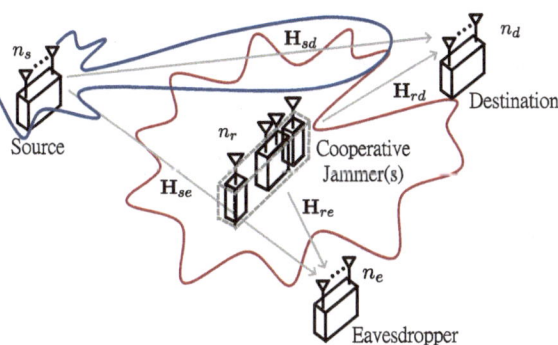

Fig. 4.4 Illustration of relays acting as pure cooperative jammers

destination and the eavesdropper are given by $\mathbf{I}_{n_d}+\mathbf{H}_{rd}\mathbf{K_a}\mathbf{H}_{rd}^H$ and $\mathbf{I}_{n_e}+\mathbf{H}_{re}\mathbf{K_a}\mathbf{H}_{re}^H$, respectively, and, thus, the achievable secrecy rate is given by

$$R_{\text{CJ}}(\mathbf{K_{x_s}}, \mathbf{K_a}) = I(\mathbf{x}_s; \mathbf{y}_d) - I(\mathbf{x}_s; \mathbf{y}_e)$$

$$= \log \frac{\det\left[\mathbf{I}_{n_s}+\mathbf{K_{x_s}}\mathbf{H}_{sd}^H(\mathbf{I}_{n_d}+\mathbf{H}_{rd}\mathbf{K_a}\mathbf{H}_{rd}^H)^{-1}\mathbf{H}_{sd}\right]}{\det\left[\mathbf{I}_{n_s}+\mathbf{K_{x_s}}\mathbf{H}_{se}^H(\mathbf{I}_{n_e}+\mathbf{H}_{re}\mathbf{K_a}\mathbf{H}_{re}^H)^{-1}\mathbf{H}_{se}\right]}. \tag{4.74}$$

The input and AN covariance matrices can then be determined by maximizing the achievable secrecy rate under total or individual power constraints. However, the optimal solution is generally difficult to find and, thus, several suboptimal approaches have been considered in the literature.

Let us consider, for example, the approach taken in [13] where the input covariance matrix is first chosen under a power covariance constraint $\mathbf{K_{x_s}} \preceq \mathbf{S}$ while assuming that no cooperative jammer exists (i.e., $\mathbf{K_a} = \mathbf{0}$). Then, the AN covariance matrix $\mathbf{K_a}$ is chosen to guarantee that there is no decrease in mutual information between the source and the destination. The latter can be viewed as a more generalized form of the zero-forcing (ZF) constraint.

In particular, with the absence of the cooperative jammer, the search for the optimal secrecy precoder $\mathbf{F}_s$ (or, equivalently, the optimal input covariance matrix $\mathbf{K_{x_s}}$) becomes equivalent to the case discussed in Sect. 3.1.2 and can be found by applying the generalized eigenvalue decomposition on the matrices $\mathbf{I}_{n_s}+(\mathbf{S}^{\frac{1}{2}})^H\mathbf{H}_{se}^H\mathbf{H}_{se}\mathbf{S}^{\frac{1}{2}}$ and $\mathbf{I}_{n_s}+(\mathbf{S}^{\frac{1}{2}})^H\mathbf{H}_{sd}^H\mathbf{H}_{sd}\mathbf{S}^{\frac{1}{2}}$. More specifically, let $\mathbf{C}$ be an invertible generalized eigenvector matrix of the two symmetric positive definite matrices $\mathbf{I}_{n_s}+(\mathbf{S}^{\frac{1}{2}})^H\mathbf{H}_{se}^H\mathbf{H}_{se}\mathbf{S}^{\frac{1}{2}}$ and $\mathbf{I}_{n_s}+(\mathbf{S}^{\frac{1}{2}})^H\mathbf{H}_{sd}^H\mathbf{H}_{sd}\mathbf{S}^{\frac{1}{2}}$ such that

$$\mathbf{C}^H\left(\mathbf{I}_{n_s}+(\mathbf{S}^{\frac{1}{2}})^H\mathbf{H}_{se}^H\mathbf{H}_{se}\mathbf{S}^{\frac{1}{2}}\right)\mathbf{C} = \mathbf{I}_{n_s} \tag{4.75}$$

and

$$\mathbf{C}^H\left(\mathbf{I}_{n_s}+(\mathbf{S}^{\frac{1}{2}})^H\mathbf{H}_{sd}^H\mathbf{H}_{sd}\mathbf{S}^{\frac{1}{2}}\right)\mathbf{C} = \mathbf{\Lambda}_d, \tag{4.76}$$

where $\mathbf{\Lambda}_d = \text{diag}\left(\lambda_1, \cdots, \lambda_{n_s}\right)$ is a positive definite diagonal matrix. Let b be the number of diagonal elements in $\mathbf{\Lambda}_d$ that are greater than 1, i.e., $\lambda_1 \geq \cdots \geq \lambda_b \geq 1 \geq \cdots \geq \lambda_{n_s}$. Let $\mathbf{\Lambda}_{d1} = \text{diag}(\lambda_1, \cdots, \lambda_b)$ and $\mathbf{\Lambda}_{d2} = \text{diag}(\lambda_{b+1}, \cdots, \lambda_{n_s})$ such that

$$\mathbf{\Lambda}_d = \begin{bmatrix} \mathbf{\Lambda}_{d1} & \mathbf{0} \\ \mathbf{0} & \mathbf{\Lambda}_{d2} \end{bmatrix}$$

and let $\mathbf{C} = [\mathbf{C}_1\ \mathbf{C}_2]$ with $\mathbf{C}_1$ and $\mathbf{C}_2$ being the $n_s \times b$ and $n_s \times (n_s - b)$ submatrices of $\mathbf{C}$. Following the derivations in Sect. 3.1.2, the optimal input covariance matrix for the case where $\mathbf{K_a} = \mathbf{0}$ can be computed as

$$\mathbf{K}^{\star}_{\mathbf{x}_s | \mathbf{K}_\mathbf{a}=0} = \mathbf{F}_s \mathbf{F}^H_s = \mathbf{S}^{\frac{1}{2}} \mathbf{C} \begin{pmatrix} \left(\mathbf{C}^H_1 \mathbf{C}_1\right)^{-1} & \mathbf{0} \\ \mathbf{0} & \mathbf{0} \end{pmatrix} \mathbf{C}^H (\mathbf{S}^{\frac{1}{2}})^H, \tag{4.77}$$

where $\mathbf{C} = [\mathbf{C}_1 \ \mathbf{C}_2]$ with $\mathbf{C}_1$ being an $n_s \times b$ submatrix. Note that $\mathbf{K}^{\star}_{\mathbf{x}_s | \mathbf{K}_\mathbf{a}=0}$ satisfies the power covariance constraint $\mathbf{0} \preceq \mathbf{K}^{\star}_{\mathbf{x}_s | \mathbf{K}_\mathbf{a}=0} \preceq \mathbf{S}$.

Given the input covariance matrix $\mathbf{K}^{\star}_{\mathbf{x}_s | \mathbf{K}_\mathbf{a}=0}$, the AN covariance matrix can then be found under the constraint that there is no decrease in the mutual information between the source and the destination. This generalized ZF constraint can be expressed as

$$\log \det \left[\mathbf{I}_{n_s} + \mathbf{K}^{\star}_{\mathbf{x}_s | \mathbf{K}_\mathbf{a}=0} \mathbf{H}^H_{sd} (\mathbf{I}_{n_d} + \mathbf{H}_{rd} \mathbf{K}_\mathbf{a} \mathbf{H}^H_{rd})^{-1} \mathbf{H}_{sd}\right] \tag{4.78}$$

$$= \log \det \left(\mathbf{I}_{n_s} + \mathbf{K}^{\star}_{\mathbf{x}_s | \mathbf{K}_\mathbf{a}=0} \mathbf{H}^H_{sd} \mathbf{H}_{sd}\right) \tag{4.79}$$

$$= \log \det \left(\left(\mathbf{C}^H_1 \mathbf{C}_1\right)^{-1} \mathbf{\Lambda}_{d1}\right) \tag{4.80}$$

where the last equality follows from (3.24). Notice that AN need not be placed in the null space of $\mathbf{H}_{rd}$ in order to satisfy the generalized ZF constraint in (4.78).

Let us consider the case where n_r is greater than the number of signal dimensions k (and refer to [13] for more general scenarios). To satisfy the generalized ZF constraint, let us consider an AN covariance matrix that can be expressed as

$$\mathbf{K}_\mathbf{a} = \mathbf{\Gamma}_\mathbf{a} \mathbf{\Pi}_\mathbf{a} \mathbf{\Gamma}^H_\mathbf{a}, \tag{4.81}$$

where $\mathbf{\Pi}_\mathbf{a} \in \mathbb{C}^{d \times d}$ is a positive definite matrix, $\mathbf{\Gamma}_\mathbf{a} \in \mathbb{C}^{n_r \times d}$, and d is the number of AN dimensions. By choosing d to be the dimension of the null space of $\mathbf{K}^{\star}_{\mathbf{x}_s | \mathbf{K}_\mathbf{a}=0} \mathbf{H}^H_{sd} \mathbf{H}_{rd}$ and by choosing $\mathbf{\Gamma}_\mathbf{a}$ to be the d right singular vectors of the null space, it follows that $\mathbf{K}^{\star}_{\mathbf{x}_s | \mathbf{K}_\mathbf{a}=0} \mathbf{H}^H_{sd} \mathbf{H}_{rd} \mathbf{\Gamma}_\mathbf{a} = \mathbf{0}$ and, thus,

$$\log \det \left[\mathbf{I}_{n_s} + \mathbf{K}^{\star}_{\mathbf{x}_s | \mathbf{K}_\mathbf{a}=0} \mathbf{H}^H_{sd} (\mathbf{I}_{n_d} + \mathbf{H}_{rd} \mathbf{K}_\mathbf{a} \mathbf{H}^H_{rd})^{-1} \mathbf{H}_{sd}\right] \tag{4.82}$$

$$= \log \det \left\{\mathbf{I}_{n_s} + \mathbf{K}^{\star}_{\mathbf{x}_s | \mathbf{K}_\mathbf{a}=0} \mathbf{H}^H_{sd}\right.$$
$$\left. \cdot \left[\mathbf{I}_{n_d} - \mathbf{H}_{rd} \mathbf{\Gamma}_\mathbf{a} (\mathbf{\Pi}^{-1}_\mathbf{a} + \mathbf{\Gamma}^H_\mathbf{a} \mathbf{H}^H_{rd} \mathbf{H}_{rd} \mathbf{\Gamma}_\mathbf{a})^{-1} \mathbf{\Gamma}^H_\mathbf{a} \mathbf{H}^H_{rd}\right] \mathbf{H}_{sd}\right\} \tag{4.83}$$

$$= \log \det \left(\mathbf{I}_{n_s} + \mathbf{K}^{\star}_{\mathbf{x}_s | \mathbf{K}_\mathbf{a}=0} \mathbf{H}^H_{sd} \mathbf{H}_{sd}\right), \tag{4.84}$$

where the first equality is obtained by substituting $\mathbf{K}_\mathbf{a}$ with the expression in (4.81) and by applying the matrix inversion lemma. This shows that the generalized ZF constraint, as defined in (4.80), is satisfied with the choice of $\mathbf{\Gamma}_\mathbf{a}$. Consequently, to maximize the achievable secrecy rate in (4.74), the matrix $\mathbf{\Pi}_\mathbf{a}$ can be chosen to minimize the logarithm of the denominator in (4.74) since the numerator no longer depends on $\mathbf{K}_\mathbf{a}$ (and, thus, $\mathbf{\Pi}_\mathbf{a}$). The logarithm of the denominator in (4.74) can be written as

$$\log \det \left[\mathbf{I}_{n_s} + \mathbf{K}^{\star}_{\mathbf{x}_s | \mathbf{K_a} = \mathbf{0}} \mathbf{H}^H_{se} (\mathbf{I}_{n_e} + \mathbf{H}_{re} \mathbf{K_a} \mathbf{H}^H_{re})^{-1} \mathbf{H}_{se} \right]$$

$$= \log \det \left(\mathbf{I}_{n_e} + \mathbf{H}_{re} \mathbf{K_a} \mathbf{H}^H_{re} + \mathbf{H}_{se} \mathbf{K}^{\star}_{\mathbf{x}_s | \mathbf{K_a} = \mathbf{0}} \mathbf{H}^H_{se} \right) - \log \det \left(\mathbf{I}_{n_e} + \mathbf{H}_{re} \mathbf{K_a} \mathbf{H}^H_{re} \right).$$

By assuming that the information leakage (i.e., $\mathbf{H}_{se} \mathbf{K}^{\star}_{\mathbf{x}_s | \mathbf{K_a} = \mathbf{0}} \mathbf{H}^H_{se}$) is large compared to the AN interference (i.e., $\mathbf{H}_{re} \mathbf{K_a} \mathbf{H}^H_{re}$) at the eavesdropper, the problem can be approximated as one that maximizes the second term above. The optimization problem can thus be formulated as

$$\min_{\mathbf{\Pi_a}} \quad \log \det \left(\mathbf{I}_{n_e} + \mathbf{H}_{re} \mathbf{\Gamma_a} \mathbf{\Pi_a} \mathbf{\Gamma}^H_{\mathbf{a}} \mathbf{H}^H_{re} \right) \tag{4.85a}$$

$$\text{subject to} \quad \text{tr}(\mathbf{\Pi_a}) \leq \bar{P}_r. \tag{4.85b}$$

Let us take the eigenvalue decomposition of $\mathbf{\Gamma}^H_{\mathbf{a}} \mathbf{H}^H_{re} \mathbf{H}_{re} \mathbf{\Gamma_a}$ as

$$\mathbf{\Gamma}^H_{\mathbf{a}} \mathbf{H}^H_{re} \mathbf{H}_{re} \mathbf{\Gamma_a} = \mathbf{U} \mathbf{D} \mathbf{U}^H \tag{4.86}$$

where $\mathbf{U} \in \mathbb{C}^{d \times d}$ is unitary and $\mathbf{D} \in \mathbb{C}^{d \times d}$ is diagonal. The optimization problem reduces to

$$\min_{\mathbf{\Pi_a}} \quad \log \det (\mathbf{I}_d + \mathbf{\Pi_a} \mathbf{U} \mathbf{D} \mathbf{U}^H) \tag{4.87}$$

$$\text{subject to} \quad \text{tr}(\mathbf{\Pi_a}) \leq \bar{P}_r. \tag{4.88}$$

Hence, the solution is given by $\mathbf{\Pi_a} = \mathbf{U} \mathbf{\Delta} \mathbf{U}^H$ with $\mathbf{\Delta}$ given by the waterfilling solution $\mathbf{\Delta} = [\eta \mathbf{I} - \mathbf{D}^{-1}]^+$ [13], where η is chosen such that $\text{tr}(\mathbf{\Pi_a}) = \text{tr}(\mathbf{\Delta}) = \bar{P}_r$.

To illustrate this idea, let us consider the simple case where $n_s = n_d = n_e = 1$, $d = 1$, and $n_r > 1$. In this case, $\mathbf{\Gamma_a} \in \mathbb{C}^{n_r \times 1}$ should be chosen to lie in the null space of $\mathbf{h}_{rd}$. Since the channel $\mathbf{h}_{re}$ is assumed to be known, $\mathbf{\Gamma_a}$ can in fact be chosen as the projection of $\mathbf{h}_{re}$ in the null space of $\mathbf{h}_{rd}$, i.e.,

$$\mathbf{\Gamma_a} = \frac{\left(\mathbf{I} - \frac{\mathbf{h}^H_{rd} \mathbf{h}_{rd}}{\|\mathbf{h}_{rd}\|^2} \right) \mathbf{h}^H_{re}}{\left\| \left(\mathbf{I} - \frac{\mathbf{h}^H_{rd} \mathbf{h}_{rd}}{\|\mathbf{h}_{rd}\|^2} \right) \mathbf{h}^H_{re} \right\|}. \tag{4.89}$$

This choice maximizes the effect of AN on the eavesdropper. If $\mathbf{h}_{re}$ is unknown, one can substitute $\mathbf{h}_{re}$ in (4.89) with $\mathbf{1}$, in which case, the resulting noise covariance matrix will be isotropic in the null space of $\mathbf{h}_{rd}$. The AN covariance matrix is then given by $\mathbf{K_a} = \bar{P}_r \mathbf{\Gamma_a} \mathbf{\Gamma}^H_{\mathbf{a}}$.

Alternatively, let us consider an example where the system consists of n_r single-antenna relays (instead of a single relay with n_r antennas). Here, we assume that the source, the destination, and the eavesdropper are each equipped with only a single antenna as well. The AN signals are assumed to be generated independently at each relay. The AN covariance matrix should be diagonal and can be written as $\mathbf{K_a} = \text{diag}(\sigma^2_{a,1}, \ldots, \sigma^2_{a,n_r})$, where $\sigma^2_{a,i}$ is the AN variance at relay i. In this case, AN cannot be intentionally nulled out at the destination. The achievable secrecy rate can be written as

$$R_{CJ} = \log\left(1 + \frac{P_s|h_{sd}|^2}{1 + \sum_{i=1}^{n_r}|h_{r_i,d}|^2\sigma^2_{a,i}}\right) - \log\left(1 + \frac{P_s|h_{se}|^2}{1 + \sum_{i=1}^{n_r}|h_{r_i,e}|^2\sigma^2_{a,i}}\right). \quad (4.90)$$

Let $\mathcal{I} = \{i : |h_{r_i,e}|^2 > |h_{r_i,d}|^2\}$ be the indices of relays that yield a better channel to the eavesdropper than to the destination and let the AN variances can be chosen as

$$\sigma^2_{a,i} = \begin{cases} \sigma^2_a, & \text{if } |h_{r_i,e}|^2 > |h_{r_i,d}|^2 \\ 0, & \text{otherwise.} \end{cases} \quad (4.91)$$

In this case, the achievable secrecy rate can be computed as

$$R_{CJ} = \log\left(1 + \frac{P_s|h_{sd}|^2}{1 + \sigma^2_a\sum_{i\in\mathcal{I}}|h_{r_i,d}|^2}\right) - \log\left(1 + \frac{P_s|h_{se}|^2}{1 + \sigma^2_a\sum_{i\in\mathcal{I}}|h_{r_i,e}|^2}\right). \quad (4.92)$$

The optimal AN variance σ^2_a can then be performed based on a simple line search. By taking the derivative of the achievable secrecy rate at $\sigma^2_a = 0$, it can be shown that, when

$$\frac{1 + 1/P_s|h_{sd}|^2}{1 + 1/P_s|h_{se}|^2} > \frac{\sum_{i\in\mathcal{I}}|h_{r_i,d}|^2}{\sum_{i\in\mathcal{I}}|h_{r_i,e}|^2},$$

a higher secrecy rate can be achieved by choosing $\sigma^2_a > 0$.

Example 4.2 The secrecy rate performance in the case of trusted relays is compared in Fig. 4.5 with respect to the relay wiretap channel in Fig. 4.1, where $n_s = n_d = n_e = 1$ and $n_r = 2$. All channel coefficients are assumed to be i.i.d. $\mathcal{CN}(0, 1)$ and the path loss exponent is $\alpha = 2$. The source, destination, and eavesdropper locations are set as $(0, 0)$, $(1, 0)$, and $(1.25, 0)$. The achievable secrecy rate of four different schemes (namely, SISOSE with no relay assistance, SISOSE with two DF relays, SISOSE with two AF relays, and SISOSE with two cooperative jammers) are shown in Fig. 4.5a under varying relay positions from $(0, 0.02)$ to $(2, 0.02)$ and a fixed transmit SNR of 10 dB. The total power constraint among source and relays are adopted. One can observe that the DF scheme performs better than the AF scheme when the relay is close to the source, and vice versa, when the relay is close to the destination. Moreover, the achievable secrecy rate of the cooperative jamming scheme increases monotonically with the decrease in its distance to the eavesdropper. The achievable secrecy rate is shown in Fig. 4.5b with fixed relay position at $(0.5, 0.02)$ and with varying transmit SNR values. One can see that the performance of the cooperative jamming (CJ) scheme improves rapidly as the SNR increases. This is due to relays' ability to completely null out the interference toward the destination when emitting the jamming signal. However, performance improvements in DF and AF schemes are limited due to constraints in the Phase-1 transmission.

4.3.2 Relays with Both Information and Artificial Noise Transmissions

In the previous section, we considered the case where the relays act as pure cooperative jammers and do not help forward the source's message. In this section, we further consider the scenario where the relay is allowed to not only emit AN signals to jam the eavesdropper's reception but also help to forward information for the source. In this case, a two-phase transmission is again needed for the source to convey information to the relay (or relays), as illustrated in Fig. 4.6. Here, both the source and the destination are also allowed to emit AN signals whenever they are not receiving.

Let us consider a relay wiretap channel that consists of a source, a relay, a destination, and an eavesdropper, which are equipped with n_s, n_r, n_d, and n_e antennas, respectively. In Phase 1, the source emits a signal that consists of both the information-bearing signal and AN. The signal can be written as

$$\mathbf{x}_s^{(1)} = \mathbf{s} + \mathbf{a}_s^{(1)}, \tag{4.93}$$

where $\mathbf{s} \in \mathbb{C}^{n_s \times 1}$ is the information-bearing signal and $\mathbf{a}_s^{(1)} \in \mathbb{C}^{n_s \times 1}$ is the jamming signal emitted by the source. The destination also emits an AN signal $\mathbf{a}_d \in \mathbb{C}^{n_d \times 1}$ in Phase 1. In this case, the signal received at the relay and the eavesdropper will be the superposition of these signals and can be written, respectively, as

$$\mathbf{y}_r = \mathbf{H}_{sr}(\mathbf{s} + \mathbf{a}_s^{(1)}) + \mathbf{H}_{dr}\mathbf{a}_d + \mathbf{w}_r, \tag{4.94a}$$

$$\mathbf{y}_e^{(1)} = \mathbf{H}_{se}(\mathbf{s} + \mathbf{a}_s^{(1)}) + \mathbf{H}_{de}\mathbf{a}_d + \mathbf{w}_e^{(1)}, \tag{4.94b}$$

where $\mathbf{H}_{sr}$ and $\mathbf{H}_{se}$ are defined as before, and $\mathbf{H}_{dr} \in \mathbb{C}^{n_r \times n_d}$ and $\mathbf{H}_{de} \in \mathbb{C}^{n_e \times n_d}$ are the channel matrices from the destination to the relay and the eavesdropper, respectively. Since the destination emits AN signals in Phase 1, it is not allowed to receive in this phase due to the half-duplex constraint.

In Phase 2, the signal vectors $\mathbf{x}_s^{(2)}$ and $\mathbf{x}_r$ are emitted by the source and the relay, respectively. The signal transmitted by the source in Phase 2 is assumed to contain only AN signals, *i.e.*, $\mathbf{x}_s^{(2)} = \mathbf{a}_s^{(2)}$, (as in [11]) whereas the signal transmitted by the relay may contain both information and AN signals, *i.e.*, $\mathbf{x}_r = \mathbf{s}_r + \mathbf{a}_r$. The received signals at the destination and the eavesdropper can be written, respectively, as

$$\mathbf{y}_d = \mathbf{H}_{sd}\mathbf{a}_s^{(2)} + \mathbf{H}_{rd}(\mathbf{s}_r + \mathbf{a}_r) + \mathbf{w}_d, \tag{4.95a}$$

$$\mathbf{y}_e^{(2)} = \mathbf{H}_{se}\mathbf{a}_s^{(2)} + \mathbf{H}_{re}(\mathbf{s}_r + \mathbf{a}_r) + \mathbf{w}_e^{(2)}. \tag{4.95b}$$

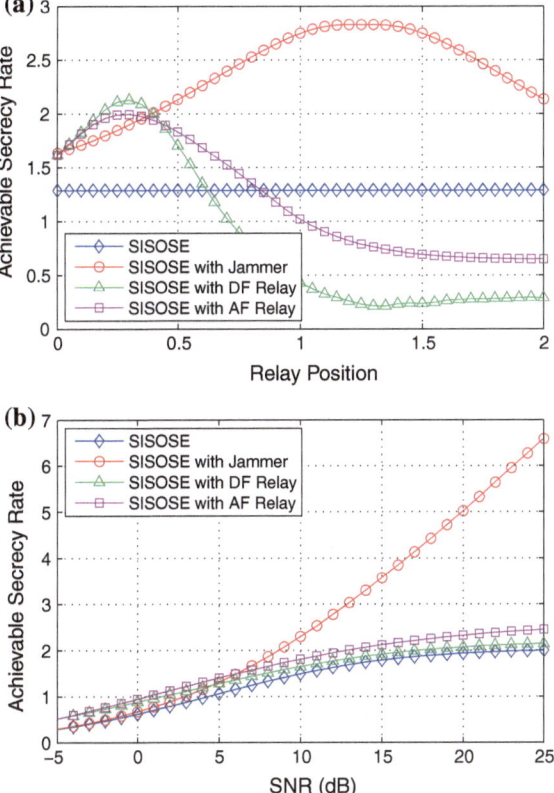

Fig. 4.5 Achievable secrecy rate for AF, DF, and CJ schemes with a trusted relay. Here, the number of antennas at the terminals are $n_s = n_d = n_e = 1$ and $n_r = 2$). **a** Varying relay position with fixed SNR, **b** Varying SNR with fixed relay position

In the above, the information-bearing signals $\mathbf{s}$ and $\mathbf{s}_r$, and the AN signals $\mathbf{a}_s^{(1)}$, $\mathbf{a}_s^{(2)}$, $\mathbf{a}_r$, and $\mathbf{a}_d$ can be designed to maximize the secrecy rate between the source and the destination. This problem can be considered for both DF and AF systems. In DF systems, the relay attempts to decode the source's message and emits a newly encoded codeword to the destination. With the addition of AN, the signal transmitted by the relay can be written as $\mathbf{x}_r = \mathbf{F}_r \mathbf{u}_r + \mathbf{a}_r$, where $\mathbf{F}_r \in \mathbb{C}^{n_r \times k_r}$ is the relay precoder, $\mathbf{u}_r \in \mathbb{C}^{k_r \times 1}$ is the newly encoded symbol vector at the relay, k_r is the number of signal dimensions at the relay, and $\mathbf{a}_r$ is the AN signal. In AF systems, the relay does not explicitly decode the source's message, but transmits a linearly amplified version of the received signal to the destination. Again, with the addition of AN, the signal transmitted by the relay can be written as $\mathbf{x}_r = \mathbf{F}_r \mathbf{y}_r + \mathbf{a}_r$, where $\mathbf{F}_r$ is the $n_r \times n_r$ relay precoder. The relay precoder should be designed such that the reception of the information-bearing signal is enhanced and the AN interference is mitigated at the destination. As an example of AF relaying, we discuss in the following a precoder

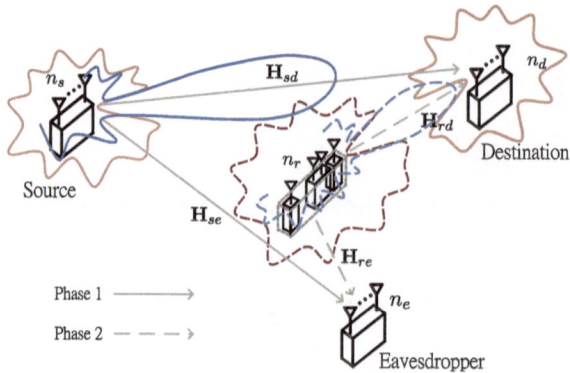

Fig. 4.6 Illustration of general distributed secrecy beamforming with artificial noise

design previously proposed in [11]. For an example on DF relaying, readers are referred to [12], where a precoder design was proposed based on the generalized singular value decomposition (GSVD) method.

In particular, let us consider an AF relay system with $n_s = n_d = n_e = M$ and $n_r = N$. Following the method proposed in [11], the information-bearing signal transmitted by the source in Phase 1 can be expressed as $\mathbf{s} = \mathbf{F}_s \mathbf{u}$, where $\mathbf{F}_s \in \mathbb{C}^{M \times k}$ is the source precoder, $\mathbf{u} \sim \mathcal{CN}(\mathbf{0}, \mathbf{I}_k)$ is the encoded symbol vector, and k is the number of signal dimensions. We assume that $k < \min\{M, N\}$. Moreover, the AN signals emitted by the source and the destination in Phase 1 can be written as $\mathbf{a}_s^{(1)} = \mathbf{G}_s^{(1)} \tilde{\mathbf{a}}_s^{(1)}$ and $\mathbf{a}_d = \mathbf{G}_d \tilde{\mathbf{a}}_d$, respectively, where $\mathbf{G}_s^{(1)} \in \mathbb{C}^{M \times (M-k)}$ and $\mathbf{G}_d \in \mathbb{C}^{M \times M}$ are the noise precoding matrices at the source and the destination, and $\tilde{\mathbf{a}}_s^{(1)} \sim \mathcal{CN}(\mathbf{0}, \mathbf{I}_{M-k})$ and $\tilde{\mathbf{a}}_d \sim \mathcal{CN}(\mathbf{0}, \mathbf{I}_M)$ are the artificially generated white noise. By substituting the above into (4.94), the received signal at the relay and the eavesdropper in Phase 1 can be written as

$$\mathbf{y}_r = \mathbf{H}_{sr} \mathbf{F}_s \mathbf{u} + \mathbf{H}_{sr} \mathbf{G}_s^{(1)} \tilde{\mathbf{a}}_s^{(1)} + \mathbf{H}_{dr} \mathbf{G}_d \tilde{\mathbf{a}}_d + \mathbf{w}_r, \qquad (4.96a)$$

$$\mathbf{y}_e^{(1)} = \mathbf{H}_{se} \mathbf{F}_s \mathbf{u} + \mathbf{H}_{se} \mathbf{G}_s^{(1)} \tilde{\mathbf{a}}_s^{(1)} + \mathbf{H}_{de} \mathbf{G}_d \tilde{\mathbf{a}}_d + \mathbf{w}_e^{(1)}. \qquad (4.96b)$$

The signal received at the relay is a mixture of the information-bearing signal and the AN signals from both the source and the destination. Notice that, since the AN vector $\tilde{\mathbf{a}}_d$ is generated by the destination, it can eventually be canceled out at the destination if it is forwarded back by the relay. Hence, the choice of $\mathbf{G}_d$ would not affect the mutual information between the source and the destination and can be chosen as $\mathbf{G}_d = \sqrt{\frac{P_d}{M}} \mathbf{I}_M$ (which is the best that one can do if the eavesdropper's channel is unknown). The design problem then lies in the choice of $\mathbf{F}_s$ and $\mathbf{G}_s^{(1)}$.

Let us take the singular value decomposition (SVD) of $\mathbf{H}_{sr}$ as

$$\mathbf{H}_{sr} = \mathbf{U}_{sr} \mathbf{\Lambda}_{sr} \mathbf{V}_{sr}^H \qquad (4.97)$$

where $\mathbf{U}_{sr} \in \mathbb{C}^{N \times N}$ and $\mathbf{V}_{sr} \in \mathbb{C}^{M \times M}$ are unitary matrices and $\mathbf{\Lambda}_{sr}$ is a diagonal matrix of singular values in descending order. To utilize the k largest singular modes of $\mathbf{H}_{sr}$ for data transmission, one can choose $\mathbf{F}_s$ as a scale of the first k columns of $\mathbf{V}_{sr}$, i.e.,

$$\mathbf{F}_s = \sqrt{\frac{P_s}{M}} \mathbf{V}_{sr}^{(1:k)} \tag{4.98}$$

where $\mathbf{V}_{sr}^{(1:k)}$ denotes the first k columns of $\mathbf{V}_{sr}$. The remaining directions, i.e., $\mathbf{V}_{sr}^{(k:M)}$, are left for AN transmissions and, thus, the noise precoder at the source is given by

$$\mathbf{G}_s^{(1)} = \sqrt{\frac{P_s}{M}} \mathbf{V}_{sr}^{(k:M)}. \tag{4.99}$$

Consequently, the source transmit power is $\text{tr}(E[\mathbf{x}_s^{(1)}(\mathbf{x}_s^{(1)})^H]) = \text{tr}(E[[(\mathbf{F}_s\mathbf{u} + \mathbf{G}_s^{(1)}\tilde{\mathbf{a}}_s^{(1)})(\mathbf{F}_s\mathbf{u} + \mathbf{G}_s^{(1)}\tilde{\mathbf{a}}_s^{(1)})^H]]) = P_s$.

In Phase 2, the AN interference from the source is first nulled out at the relay by multiplying its received signal $\mathbf{y}_r$ with a matrix $\mathbf{W} \in \mathbb{C}^{k \times N}$, which is chosen such that [11]

$$\mathbf{W}\mathbf{H}_{sr}\mathbf{G}_s^{(1)} = \mathbf{0}_{k \times (N-k)}. \tag{4.100}$$

This can be done by choosing $\mathbf{W}$ as the Hermitian of the first k columns of $\mathbf{U}_{sr}$, i.e.,

$$\mathbf{W}_r = (\mathbf{U}_{sr}^{(1:k)})^H. \tag{4.101}$$

With the choice of the signal and AN precoders given above, the effective received signal at the relay can be written as

$$\mathbf{W}_r\mathbf{y}_r = (\mathbf{U}_{sr}^{(1:k)})^H \mathbf{y}_r \tag{4.102}$$

$$= \sqrt{\frac{P_s}{M}} \mathbf{\Lambda}_{sr}^{(1:k)}\mathbf{u} + \sqrt{\frac{P_d}{M}}(\mathbf{U}_{sr}^{(1:k)})^H\mathbf{H}_{dr}\tilde{\mathbf{a}}_d + (\mathbf{U}_{sr}^{(1:k)})^H\mathbf{w}_r \tag{4.103}$$

where $\mathbf{\Lambda}_{sr}^{(1:k)} \in \mathbb{C}^{k \times k}$ is a diagonal matrix with the k largest singular values of $\mathbf{H}_{sr}$ on its diagonal. Here, we have replaced $\mathbf{G}_d$ with the identity matrix $\sqrt{\frac{P_d}{M}}\mathbf{I}_M$. Notice that the AN interference received from the destination is not explicitly dealt with since it can be canceled out at the destination.

Then, the effective received signal at the relay, i.e., $\mathbf{W}_r\mathbf{y}_r$, is multiplied by the diagonal matrix $\mathbf{D}_r$, which determines the power allocation among different signal dimensions at the relay. The resulting signal vector is then transmitted over the k largest singular modes of $\mathbf{H}_{rd}$. Specifically, let

$$\mathbf{H}_{rd} = \mathbf{U}_{rd}\mathbf{\Lambda}_{rd}\mathbf{V}_{rd}^H \tag{4.104}$$

be the SVD of $\mathbf{H}_{rd}$, where $\mathbf{U}_{rd} \in \mathbb{C}^{M \times M}$ and $\mathbf{V}_{rd} \in \mathbb{C}^{N \times N}$ are unitary matrices, and $\mathbf{\Lambda}_{rd}$ is a diagonal matrix of singular values of $\mathbf{H}_{rd}$ in descending order. Therefore, the information signal transmitted by the relay can be expressed as

$$\mathbf{s}_r = \sqrt{\frac{P_r}{N}} \mathbf{V}_{rd}^{(1:k)} \mathbf{D}_r \mathbf{W}_r \mathbf{y}_r. \tag{4.105}$$

The effective relay precoder in this case is given by

$$\mathbf{F}_r = \sqrt{\frac{P_r}{N}} \mathbf{V}_{rd}^{(1:k)} \mathbf{D}_r \mathbf{W}_r. \tag{4.106}$$

Notice that only k dimensions are utilized for information forwarding and, thus, $N - k$ dimensions remain for the transmission of AN signals. The AN signal emitted by the relay in Phase 2 can be expressed as $\mathbf{a}_r = \mathbf{G}_r \tilde{\mathbf{a}}_r$, where $\mathbf{G}_r \in \mathbb{C}^{N \times (N-k)}$ is the noise precoder and $\tilde{\mathbf{a}}_r \sim \mathcal{CN}(\mathbf{0}, \mathbf{I}_{N-k})$. In particular, the noise precoder can be chosen as

$$\mathbf{G}_r = \sqrt{\frac{P_r}{N}} \mathbf{V}_{rd}^{(k:N)} \tag{4.107}$$

so that it lies in a subspace orthogonal to that of the information signal. By assuming that an equal amount of power is allocated to the signal transmitted on each dimension, the diagonal matrix $\mathbf{D}_r$ is chosen as

$$\{\mathbf{D}_r\}_{m,m}^{-2} = \frac{P_s}{M} \lambda_{sr,m}^2 + \frac{P_d}{M} \{(\mathbf{U}_{sr}^{(1:k)})^H \mathbf{H}_{dr} \mathbf{H}_{dr}^H \mathbf{U}_{sr}^{(1:k)}\}_{m,m} + 1, \tag{4.108}$$

where $\{\mathbf{A}\}_{i,j}$ denotes the (i, j)th entry of $\mathbf{A}$ and $\lambda_{sr,m}$ is the m-th diagonal entry of $\mathbf{\Lambda}_{sr}$. Moreover, in Phase 2, the source is also allowed to emit a new AN signal, denoted by $\mathbf{a}_s^{(2)} = \mathbf{G}_s^{(2)} \tilde{\mathbf{a}}_s^{(2)}$, to help disrupt the reception at the eavesdropper.

By substituting the above into (4.95), the signal received at the destination and the eavesdropper in Phase 2 can be written as

$$\mathbf{y}_d = \mathbf{H}_{sd} \mathbf{G}_s^{(2)} \tilde{\mathbf{a}}_s^{(2)} + \mathbf{H}_{rd} \mathbf{F}_r \mathbf{y}_r + \mathbf{H}_{rd} \mathbf{G}_r \tilde{\mathbf{a}}_r + \mathbf{w}_d, \tag{4.109a}$$

$$\mathbf{y}_e^{(2)} = \mathbf{H}_{se} \mathbf{G}_s^{(2)} \tilde{\mathbf{a}}_s^{(2)} + \mathbf{H}_{re} \mathbf{F}_r \mathbf{y}_r + \mathbf{H}_{re} \mathbf{G}_r \tilde{\mathbf{a}}_r + \mathbf{w}_e^{(2)}. \tag{4.109b}$$

Recall that the AN signal emitted by the destination in Phase 1 is not nulled out by the choice of $\mathbf{W}$ and is also forwarded to the destination in Phase 2. However, the destination will be able to cancel this signal out since it is generated by itself and, thus, known a priori at the destination. Moreover, by employing the concept of interference alignment [11], the noise precoder $\mathbf{G}_s^{(2)}$ at the source can be designed such that the AN signals emitted by the source and the relay, i.e., $\mathbf{G}_s^{(2)} \tilde{\mathbf{a}}_s^{(2)}$ and $\mathbf{G}_r \tilde{\mathbf{a}}_r$, lie in the same subspace at the destination. Further details on how this can be done can be found in [11]. In this case, the destination will be able to null out the AN from both the source and the relay simultaneously by employing the receive filter $\mathbf{W}_d = (\mathbf{U}_{rd}^{(1:k)})^H$. By canceling out the self-generated AN signal, the effective received signal at the destination can be written as

$$
\begin{aligned}
\mathbf{W}_d \mathbf{y}_d &= (\mathbf{U}_{rd}^{(1:k)})^H \mathbf{y}_d \\
&= \sqrt{\frac{P_r P_s}{NM}} \boldsymbol{\Lambda}_{rd}^{(1:k)} \mathbf{D}_r \boldsymbol{\Lambda}_{sr}^{(1:k)} \mathbf{u} + \sqrt{\frac{P_r}{N}} \boldsymbol{\Lambda}_{rd}^{(1:k)} \mathbf{D}_r (\mathbf{U}_{sr}^{(1:k)})^H \mathbf{w}_r + (\mathbf{U}_{rd}^{(1:k)})^H \mathbf{w}_d.
\end{aligned}
$$

This shows that the source message has been allocated to the k largest singular modes of the two channel matrices $\mathbf{H}_{sr}$ and $\mathbf{H}_{rd}$ so that the SNR at the destination is maximized. The mutual information between the source and the destination can then be written as [11]

$$
I(\mathbf{u}; \mathbf{y}_d^{(2)}) = \sum_{m=1}^{k} \log \left(1 + \frac{P_r P_s}{NM} \frac{\lambda_{rd,m}^2 \lambda_{sr,m}^2}{\frac{P_r}{N} \lambda_{rd,m}^2 + \{\mathbf{D}_r\}_{m,m}^{-2}} \right) \tag{4.110}
$$

where $\lambda_{sr,m}$ and $\lambda_{rd,m}$ are the mth diagonal entries of $\boldsymbol{\Lambda}_{sr}$ and $\boldsymbol{\Lambda}_{rd}$, respectively.

It is interesting to note that the precoders are chosen only to null out the noise at the destination, but not at the eavesdropper. Hence, the AN signals emitted by all terminals will cause performance degradation at the eavesdropper. Let us collect the received signals at the eavesdropper in both phases into the vector

$$
\mathbf{y}_e = \begin{pmatrix} \mathbf{y}_e^{(1)} \\ \mathbf{y}_e^{(2)} \end{pmatrix} = \underbrace{\begin{pmatrix} \sqrt{\frac{P_s}{M}} \mathbf{H}_{se} \mathbf{V}_{sr}^{(1:k)} \\ \sqrt{\frac{P_r P_s}{NM}} \mathbf{H}_{re} \mathbf{V}_{rd}^{(1:k)} \mathbf{D}_r \boldsymbol{\Lambda}_{sr}^{(1:k)} \end{pmatrix}}_{\tilde{\mathbf{H}}_e} \mathbf{u} + \underbrace{\begin{pmatrix} \tilde{\mathbf{w}}_e^{(1)} \\ \tilde{\mathbf{w}}_e^{(2)} \end{pmatrix}}_{\tilde{\mathbf{w}}_e} \tag{4.111}
$$

where

$$
\tilde{\mathbf{w}}_e^{(1)} = \sqrt{\frac{P_s}{M}} \mathbf{H}_{se} \mathbf{V}_{sr}^{(k:M)} \tilde{\mathbf{a}}_s^{(1)} + \sqrt{\frac{P_d}{M}} \mathbf{H}_{de} \tilde{\mathbf{a}}_d + \mathbf{w}_e^{(1)}
$$

and

$$
\begin{aligned}
\tilde{\mathbf{w}}_e^{(2)} = {}& \mathbf{H}_{se} \mathbf{G}_s^{(2)} \tilde{\mathbf{a}}_s^{(2)} + \sqrt{\frac{P_r}{N}} \sqrt{\frac{P_d}{M}} \mathbf{H}_{re} \mathbf{V}_{rd}^{(1:k)} \mathbf{D}_r (\mathbf{U}_{sr}^{(1:k)})^H \mathbf{H}_{dr} \tilde{\mathbf{a}}_d \\
& + \sqrt{\frac{P_r}{N}} \mathbf{H}_{re} \mathbf{V}_{rd}^{(1 \cdot k)} \mathbf{D}_r (\mathbf{U}_{sr}^{(1 \cdot k)})^H \mathbf{w}_r + \sqrt{\frac{P_r}{N}} \mathbf{H}_{re} \mathbf{V}_{rd}^{(k:N)} \tilde{\mathbf{a}}_r + \mathbf{w}_e^{(2)}
\end{aligned}
$$

are the effective noise at the eavesdropper in Phases 1 and 2, respectively. As defined in (4.111), $\tilde{\mathbf{H}}_e$ and $\tilde{\mathbf{w}}_e$ can be viewed as the effective channel and noise at the eavesdropper. Hence, the mutual information between the source and the eavesdropper can be computed as [11]

$$
I(\mathbf{u}; \mathbf{y}_e) = \log \det \left(\mathbf{I} + \tilde{\mathbf{H}}_e \tilde{\mathbf{H}}_e^H \mathbf{K}_{\tilde{\mathbf{w}}_e}^{-1} \right), \tag{4.112}
$$

where

$$\mathbf{K}_{\tilde{\mathbf{w}}_e} = \mathrm{E}\left[\begin{pmatrix} \tilde{\mathbf{w}}_e^{(1)} \\ \tilde{\mathbf{w}}_e^{(2)} \end{pmatrix} ((\tilde{\mathbf{w}}_e^{(1)})^H, (\tilde{\mathbf{w}}_e^{(2)})^H)\right] \qquad (4.113)$$

is the covariance matrix of the effective noise. One can observe that, as P_s, P_r, and P_d becomes large, the thermal noise becomes negligible and the effective noise at the eavesdropper becomes dominated by AN. As a result, the mutual information between the source and the eavesdropper eventually converges to a constant since both the numerator and the denominator increase with the transmit power. In contrast, the mutual information between the source and the destination, as given in (4.110), increases without bound as the transmit power increases. Hence, the achievable secrecy rate can be made arbitrarily large by increasing the transmit power. Notice that this is achieved without knowledge of the eavesdropper's channel since the choice of precoders depends only on the channels between the source, the relay, and the destination.

4.4 Summary and Discussions

In this chapter, the secrecy beamforming and precoding techniques introduced in the previous chapter were extended to the case with cooperative relays. The multiple antennas at the relays provide additional spatial degrees of freedom that can be utilized to enhance the difference between the channels at the destination and the eavesdropper and, in this way, increase the achievable secrecy rate. However, in relay systems, further security threats caused by the additional transmission required to achieve source-relay coordination and the trustworthiness of the relay must also be carefully addressed. Most discussions in this chapter were based on the popular DF and AF relaying schemes. In these schemes, communication between the source and the destination is achieved through two phases. The transmission in both phases are susceptible to eavesdropping and must be designed to avoid information leakage to the eavesdropper. First, we considered the case where the relays are trustworthy and are utilized as a distributed multi-antenna extension of the source. Then, we considered the case where the relays are untrustworthy and are, in fact, treated as the eavesdropper in the system. Only the AF scheme is discussed in this case since the relay is not allowed to decode the message.

In addition to the distributed secrecy beamforming and precoding schemes, the relay can also assist in transmitting AN signals to jam the eavesdropper's reception. Two cases were considered: (i) the case where the relay acts as a pure cooperative jammer and (ii) the case where the relays transmit both information and AN signals simultaneously. In the first case, only a single-phase transmission is required since no information needs to be received by the relay. In the second case, a more general scenario is considered where, not only the relay, but also the source and the destination may also participate in the transmission of AN signals. In both cases with and without AN, the optimal source and relay precoders that maximize the achievable secrecy rate are difficult to find in general. However, several efficient methods have been proposed based on ZF or GSVD approaches.

It is interesting to remark that, in this chapter, we focused only on systems that consist of one destination and one eavesdropper. These discussions can also be generalized to the case with multiple destinations and eavesdroppers, as in Chap. 3. Readers are referred to [3] for further discussions on this topic. Moreover, it is also necessary to note that the discussions in this chapter are mostly based on cases where the channel matrices are known to all terminals. In practice, channel knowledge is always subject to errors either due to noise in the channel estimation process or due to limited feedback. In fact, the eavesdropper's channel may often be completely unknown. In these cases, the ergodic secrecy or the secrecy outage can be utilized as the design criterion. Readers are referred to [11] for further discussions on this issue.

References

1. Gomez-Cuba F, Asorey-Cacheda R, Gonzalez-Castano F (2012) A survey on cooperative diversity for wireless networks. IEEE Commun Surveys Tuts 14(3):822–835
2. He X, Yener A (2010) Cooperation with an untrusted relay: a secrecy perspective. IEEE Trans Inf Theory 56(8):3807–3827
3. Dong L, Han Z, Petropulu A, Poor H (2010) Improving wireless physical layer security via cooperating relays. IEEE Trans Signal Process 58(3):1875–1888
4. Zhang J, Gursoy M (2010) Collaborative relay beamforming for secrecy. In: Proceedings of IEEE International Conference on Communications (ICC)
5. Horn RA, Johnson CR (1985) Matrix analysis. Cambridge Univerity Press, Cambridge
6. Solis FJ, Wets RJ-B (1981) Minimization by random search techniques. Math Oper Res 6:19–30
7. Oohama Y (2001) Coding for relay channels with confidential messages. In: Proceedings of IEEE Information Theory, Workshop, 87–89
8. Jeong C, Kim I-M, Kim DI (2012) Joint secure beamforming design at the source and the relay for an amplify-and-forward MIMO untrusted relay system. IEEE Trans Signal Process 60(1):310–325
9. Maio AD, Huang Y, Palomar DP, Zhang S, Farina A (2011) Fractional QCQP with applications in ML steering direction estimation for radar detection. IEEE Trans Signal Process 59(1):172–185
10. Rong Y, Gao F (2009) Optimal beamforming for non-regenerative MIMO relays with direct link. IEEE Commun Lett 13(12):926–928
11. Ding Z, Peng M, Chen H-H (2012) A general relaying transmission protocol for MIMO secrecy communications. IEEE Trans Commun 60(11):3461–3471
12. Huang J, Swindlehurst AL (2011) Cooperative jamming for secure communications in MIMO relay networks. IEEE Trans Signal Process 59(10):4871–4884
13. Fakoorian S, Swindlehurst A (2011) Solutions for the MIMO Gaussian wiretap channel with a cooperative jammer. IEEE Trans Signal Process 59(10):5013–5022

Chapter 5
Secrecy-Enhancing Channel Estimation in Multi-Antenna Wireless Systems

Abstract This chapter provides an overview of the so-called discriminatory channel estimation (DCE) scheme, that is, a training and channel estimation scheme utilized to enhance physical layer secrecy in the channel estimation phase. Instead of focusing on the data transmission, as done in most works introduced in previous chapters, the DCE scheme focuses on the training signal design in the channel estimation phase and aims to discriminate the channel estimation performance at the destination and the eavesdropper. By allowing the destination to obtain a better channel estimate than the eavesdropper, the difference in signal qualities at the two receivers will increase, leaving more room for secrecy coding in the data transmission phase.

Keywords Channel estimation · Multiple-input multiple-output (MIMO) · Pilot signal · Artificial noise · Two-way training · Secrecy

In previous sections, signal processing techniques in the data transmission phase, e.g., secrecy beamforming and precoding schemes (with and without AN), were introduced and were shown to be effective in terms of enlarging the difference in signal quality at the destination and the eavesdropper. Interestingly, the desired signal quality difference can be achieved not only in the data transmission phase but also in the channel estimation phase through proper design of the training and channel estimation scheme. In fact, poor channel estimation quality results in low signal reception quality during data transmission [1, 2]. Hence, by discriminating the channel estimation performance at the destination and the eavesdropper, the difference in signal qualities at the two receivers in the data transmission phase will also be enhanced, leaving more room for secrecy coding. This has motivated the design of the so-called *discriminatory channel estimation* (DCE) scheme proposed in [3, 4]. Since the DCE scheme is employed in the channel estimation phase, it does not limit the use of signal processing techniques in the data transmission phase. Hence, most techniques described in previous chapters can also be employed with the DCE scheme to further enhance secrecy in the physical layer.

In conventional training schemes [5], pure pilot signals are emitted by the source to facilitate channel estimation at all receivers. This technique is sufficient in most conventional scenarios, where there is typically no need to differentiate the performance at different receivers. However, in secrecy applications, it is desirable to degrade the channel estimation performance at the eavesdropper or unauthorized receiver, while maintaining reliable performance at the intended destination. This is achieved in the DCE scheme by inserting artificial noise (AN) into the training signal to mask the transmission of the pilot matrix. This is similar to the use of AN or jamming signals in the data transmission phase, as described in previous chapters. However, different from that in the data transmission phase, channel knowledge is not yet available at the source, and thus it is difficult to place AN in an appropriately chosen subspace to avoid interference at the destination. Hence, a multi-stage operation is needed to first provide the source with a preliminary estimate of the main channel and to then transmit training signals with AN embedded in the null space of the estimated main channel to impair the channel estimation performance at the eavesdropper. However, it is challenging to provide the source with preliminary knowledge of the channel without benefiting the channel estimation at the eavesdropper. To achieve this task, two DCE schemes were proposed in the literature, namely the feedback-and-retraining [3] and the two-way training [4] DCE schemes. These two schemes are described in the following sections.

5.1 Feedback-and-Retraining Discriminatory Channel Estimation Scheme

In this section, we introduce the feedback-and-retraining DCE scheme previously proposed in [3]. In particular, a basic feedback-and-retraining DCE scheme consists of two stages, i.e., a preliminary training stage and a feedback-and-retraining stage. In the preliminary training stage, a training signal that consists of a conventional pilot matrix is first emitted by the source to enable a rough channel estimation at the destination. Then, in the feedback-and-retraining stage, the destination sends this rough estimate to the source, who then transmits a new training signal with AN embedded in the null space of the estimated channel to disrupt the reception at the eavesdropper.

Specifically, let us consider a multiple-input multiple-output multiantenna eavesdropper (MIMOME) system, as shown in Fig. 5.1, where the source, the destination and the eavesdropper are equipped with n_s, n_d, and n_e antennas, respectively. Here, we assume that the number of antennas at the source is greater than that at the destination, i.e., $n_s > n_d$. In the preliminary training stage (i.e., stage 0), the source first emits an $n_s \times T_0$ training signal

$$\mathbf{X}_0 = \sqrt{\frac{P_0 T_0}{n_s}} \mathbf{C}_0, \tag{5.1}$$

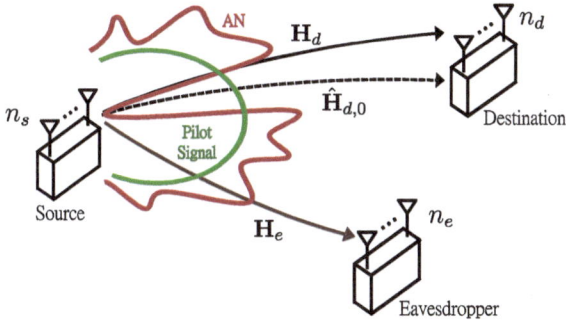

Fig. 5.1 Illustration of AN-assisted training in DCE schemes

where P_0 is the pilot signal power, T_0 is the training length, and $\mathbf{C}_0 \in \mathbb{C}^{n_s \times T_0}$ is a semi-unitary pilot matrix satisfying $\mathbf{C}_0\mathbf{C}_0^H = \mathbf{I}_{n_s}$. Each row of $\mathbf{X}_0$ represents a training signal vector sent over one of the n_s transmit antennas. The received signals at the destination and the eavesdropper are given by

$$\mathbf{Y}_0 = \mathbf{H}_d\mathbf{X}_0 + \mathbf{W}_0, \tag{5.2}$$

$$\mathbf{Z}_0 = \mathbf{H}_e\mathbf{X}_0 + \mathbf{V}_0, \tag{5.3}$$

where $\mathbf{H}_d \in \mathbb{C}^{n_d \times n_s}$ and $\mathbf{H}_e \in \mathbb{C}^{n_e \times n_s}$ are the main and the eavesdropper channel matrices, and $\mathbf{W}_0 \in \mathbb{C}^{n_d \times T_0}$ and $\mathbf{V}_0 \in \mathbb{C}^{n_e \times T_0}$ are the additive white Gaussian noise (AWGN) matrices at the destination and the eavesdropper, respectively. The entries of $\mathbf{H}_d$, $\mathbf{H}_e$, $\mathbf{W}_0$, and $\mathbf{V}_0$ are i.i.d. Gaussian with 0 mean and variances $\sigma_{h_d}^2$, $\sigma_{h_e}^2$, σ_w^2 and σ_v^2, respectively.

In the preliminary training stage, the destination can compute a preliminary estimate of the channel matrix $\mathbf{H}_d$ based on its received signal $\mathbf{Y}_0$. By adopting the linear minimum mean square error (LMMSE) estimator [6], the estimate of the channel matrix is given by

$$\hat{\mathbf{H}}_{d,0} = \sigma_{h_d}^2\mathbf{Y}_0\left(\sigma_{h_d}^2\mathbf{X}_0^H\mathbf{X}_0 + \sigma_w^2\mathbf{I}_{T_0}\right)^{-1}\mathbf{X}_0^H, \tag{5.4}$$

where $\mathbf{I}_{T_0}$ is the $T_0 \times T_0$ identity matrix. The channel estimation error can be defined as $\Delta\mathbf{H}_{d,0} \triangleq \hat{\mathbf{H}}_{d,0} - \mathbf{H}_d$, whose correlation can be computed as

$$\mathrm{E}\left[\Delta\mathbf{H}_{d,0}^H\Delta\mathbf{H}_{d,0}\right] = n_d\left(\frac{1}{\sigma_{h_d}^2}\mathbf{I}_{n_s} + \frac{P_0T_0}{n_s\sigma_w^2}\mathbf{C}_0\mathbf{C}_0^H\right)^{-1} \tag{5.5}$$

$$= n_d\left(\frac{1}{\sigma_{h_d}^2} + \frac{P_0T_0}{n_s\sigma_w^2}\right)^{-1}\mathbf{I}_{n_s}. \tag{5.6}$$

The channel estimation performance can be measured by the normalized mean square error (NMSE) given by

$$\text{NMSE}_d^{(0)} \triangleq \frac{\text{tr}\left(\text{E}\left[\Delta \mathbf{H}_{d,0}^H \Delta \mathbf{H}_{d,0}\right]\right)}{n_s n_d} = \left(\frac{1}{\sigma_{h_d}^2} + \frac{P_0 T_0}{n_s \sigma_w^2}\right)^{-1}, \qquad (5.7)$$

i.e., the MSE normalized by the number of estimating parameters.

Notice that, in the preliminary training stage, the training signal contains only the pilot matrix, and thus provides equal benefits to the channel estimation at the eavesdropper as well. Therefore, based on the received signal $\mathbf{Z}_0$ given in (5.3), the eavesdropper can also compute an LMMSE estimate of its channel in stage 0, which is given by

$$\hat{\mathbf{H}}_{e,0} = \sigma_{h_e}^2 \mathbf{Z}_0 \left(\sigma_{h_e}^2 \mathbf{X}_0^H \mathbf{X}_0 + \sigma_v^2 \mathbf{I}_{T_0}\right)^{-1} \mathbf{X}_0^H. \qquad (5.8)$$

The associated NMSE at the eavesdropper is given by

$$\text{NMSE}_e^{(0)} \triangleq \frac{\text{tr}\left(\text{E}\left[\Delta \mathbf{H}_{e,0}^H \Delta \mathbf{H}_{e,0}\right]\right)}{n_s n_e} = \left(\frac{1}{\sigma_{h_e}^2} + \frac{P_0 T_0}{n_s \sigma_v^2}\right)^{-1}. \qquad (5.9)$$

In the feedback-and-retraining stage, the destination first feeds back the channel estimate obtained in the previous stage to the source. The source then emits a new pilot signal masked by AN to provide additional training for the source while simultaneously limiting the channel estimation performance at the eavesdropper. With knowledge of the preliminary channel estimate obtained in the previous stage, the source can place AN in the null space of the estimated channel to minimize its interference at the destination. The dimension of the null space is nonzero since $n_s > n_d$. The training signal emitted in the feedback-and-retraining stage can be written as

$$\mathbf{X}_1 = \sqrt{\frac{P_1 T_1}{n_s}} \mathbf{C}_1 + \mathbf{N}_{\hat{\mathbf{H}}_{d,0}} \mathbf{A}_1, \qquad (5.10)$$

where P_1 is the pilot signal power, T_1 is the training length, $\mathbf{C}_1 \in \mathbb{C}^{n_s \times T_1}$ is a semi-unitary pilot matrix satisfying $\mathbf{C}_1 \mathbf{C}_1^H = \mathbf{I}_{n_s}$, $\mathbf{N}_{\hat{\mathbf{H}}_{d,0}} \in \mathbb{C}^{n_s \times (n_s - n_d)}$ is a matrix whose columns form an orthonormal basis for the null space of $\hat{\mathbf{H}}_{d,0}$, and $\mathbf{A}_1 \in \mathbb{C}^{(n_s - n_d) \times T_1}$ is an AN matrix with entries being i.i.d. $\mathcal{CN}(0, \sigma_{a,1}^2)$. Similarly, the received signals at the destination and the eavesdropper can be written as

$$\mathbf{Y}_1 = \mathbf{H}_d \mathbf{X}_1 + \mathbf{W}_1, \qquad (5.11)$$
$$\mathbf{Z}_1 = \mathbf{H}_e \mathbf{X}_1 + \mathbf{V}_1, \qquad (5.12)$$

where $\mathbf{W}_1 \in \mathbb{C}^{n_d \times T_1}$ and $\mathbf{V}_1 \in \mathbb{C}^{n_e \times T_1}$ are the AWGN matrices whose entries are i.i.d. Gaussian with 0 mean and variances σ_w^2 and σ_v^2, respectively.

In this stage, both the destination and the eavesdropper can refine their channel estimates using the received signals in both stages. In particular, at the destination, the signals received in stages 0 and 1 can form an effective observation matrix

$$\mathbf{Y} \triangleq [\mathbf{Y}_0, \ \mathbf{Y}_1] = \mathbf{H}_d \bar{\mathbf{C}} + \bar{\mathbf{W}}, \tag{5.13}$$

where $\bar{\mathbf{C}} \triangleq \left[\sqrt{\frac{P_0 T_0}{n_s}} \mathbf{C}_0, \ \sqrt{\frac{P_1 T_1}{n_s}} \mathbf{C}_1\right]$ and $\bar{\mathbf{W}} \triangleq \left[\mathbf{W}_0, \ \mathbf{H}_d \mathbf{N}_{\hat{\mathbf{H}}_{d,0}} \mathbf{A}_1 + \mathbf{W}_1\right]$. By taking $\mathbf{Y}$ as the observation, the LMMSE estimate can be computed as

$$\hat{\mathbf{H}}_{d,1} = n_d \sigma_{h_d}^2 \mathbf{Y} \left(n_d \sigma_{h_d}^2 \bar{\mathbf{C}}^H \bar{\mathbf{C}} + \mathbf{R}_{\bar{\mathbf{W}}}\right)^{-1} \bar{\mathbf{C}}^H, \tag{5.14}$$

where $\mathbf{R}_{\bar{\mathbf{W}}} \triangleq \mathrm{E}\left[\bar{\mathbf{W}}^H \bar{\mathbf{W}}\right]$ is the correlation matrix of $\bar{\mathbf{W}}$. By utilizing the fact that $\mathbf{W}_0$, $\mathbf{W}_1$, and $\mathbf{A}_1$ are statistically independent, and the fact that $\mathbf{H}_d \mathbf{N}_{\hat{\mathbf{H}}_{d,0}} \mathbf{A}_1 = -\Delta \mathbf{H}_{d,0} \mathbf{N}_{\hat{\mathbf{H}}_{d,0}} \mathbf{A}_1$, the correlation matrix can be written as

$$\mathbf{R}_{\bar{\mathbf{W}}} = \begin{bmatrix} n_d \sigma_w^2 \mathbf{I}_{T_0} & \mathbf{0} \\ \mathbf{0} & \left(\mathrm{E}[\|\Delta \mathbf{H}_{d,0} \mathbf{N}_{\hat{\mathbf{H}}_{d,0}}\|_F^2] \sigma_{a,1}^2 + n_d \sigma_w^2\right) \mathbf{I}_{T_1} \end{bmatrix}, \tag{5.15}$$

where $\|\cdot\|_F^2$ is the operator that yields the Frobenius norm of a matrix. Moreover, by the orthogonality principle [6], we know that $\hat{\mathbf{H}}_{d,0}$ and $\Delta \mathbf{H}_{d,0}$ are statistically uncorrelated, and thus we have

$$\mathrm{E}[\|\Delta \mathbf{H}_{d,0} \mathbf{N}_{\hat{\mathbf{H}}_{d,0}}\|_F^2] = \mathrm{tr}\left(\mathrm{E}\left[\mathbf{N}_{\hat{\mathbf{H}}_{d,0}}^H \Delta \mathbf{H}_{d,0}^H \Delta \mathbf{H}_{d,0} \mathbf{N}_{\hat{\mathbf{H}}_{d,0}}\right]\right) \tag{5.16}$$

$$= n_d (n_s - n_d) \cdot \left(\frac{1}{\sigma_{h_d}^2} + \frac{P_0 T_0}{n_s \sigma_w^2}\right)^{-1} \tag{5.17}$$

$$= n_d (n_s - n_d) \cdot \mathrm{NMSE}_d^{(0)}, \tag{5.18}$$

which follows from (5.6) and (5.7). The associated NMSE is given by

$$\mathrm{NMSE}_d^{(1)} \triangleq \frac{\mathrm{tr}\left(\mathrm{E}\left[\Delta \mathbf{H}_{d,1}^H \Delta \mathbf{H}_{d,1}\right]\right)}{n_s n_d} = \frac{\mathrm{tr}\left(\left(\frac{1}{n_d \sigma_{h_d}^2}\mathbf{I}_{n_s} + \bar{\mathbf{C}} \mathbf{R}_{\bar{\mathbf{W}}}^{-1} \bar{\mathbf{C}}^H\right)^{-1}\right)}{n_s n_d}$$

$$= \left(\frac{1}{\mathrm{NMSE}_d^{(0)}} + \frac{P_1 T_1 / n_s}{\mathrm{NMSE}_d^{(0)} \cdot (n_s - n_d)\sigma_{a,1}^2 + \sigma_w^2}\right)^{-1}, \tag{5.19}$$

Similarly, the received signals at the eavesdropper in both stages can also be collected into the matrix

$$\mathbf{Z} \triangleq [\mathbf{Z}_0, \ \mathbf{Z}_1] \triangleq \mathbf{H}_e \bar{\mathbf{C}} + \bar{\mathbf{V}}, \tag{5.20}$$

where $\bar{\mathbf{C}}$ is defined as in (5.13) and $\bar{\mathbf{V}} \triangleq \left[\mathbf{V}_0, \ \mathbf{H}_e \mathbf{N}_{\hat{\mathbf{H}}_{d,0}} \mathbf{A}_1 + \mathbf{V}_1 \right]$. By taking $\mathbf{Z}$ as the observation, the LMMSE estimate at the eavesdropper can be computed as

$$\hat{\mathbf{H}}_{e,1} = n_e \sigma_{h_e}^2 \mathbf{Z} \left(n_e \sigma_{h_e}^2 \bar{\mathbf{C}}^H \bar{\mathbf{C}} + \mathbf{R}_{\bar{\mathbf{V}}} \right)^{-1} \bar{\mathbf{C}}^H \tag{5.21}$$

where $\mathbf{R}_{\bar{\mathbf{V}}} \triangleq \mathbb{E}[\bar{\mathbf{V}}^H \bar{\mathbf{V}}]$ is the correlation matrix of $\bar{\mathbf{V}}$. Because of the independence between $\mathbf{H}_e$ and $\hat{\mathbf{H}}_{d,0}$, the correlation matrix can be computed as

$$\mathbf{R}_{\bar{\mathbf{V}}} = \begin{bmatrix} n_e \sigma_v^2 \mathbf{I}_{T_0} & \mathbf{0} \\ \mathbf{0} & n_e \left[(n_s - n_d) \sigma_{a,1}^2 \sigma_{h_e}^2 + \sigma_v^2 \right] \mathbf{I}_{T_1} \end{bmatrix}. \tag{5.22}$$

Consequently, the NMSE of the estimate $\hat{\mathbf{H}}_{e,1}$ at the eavesdropper can be written as

$$\begin{aligned} \text{NMSE}_e^{(1)} &\triangleq \frac{\operatorname{tr} \left(\mathbb{E} \left[\Delta \mathbf{H}_{e,1}^H \Delta \mathbf{H}_{e,1} \right] \right)}{n_s n_e} = \frac{\operatorname{tr} \left(\left(\frac{1}{n_e \sigma_{h_e}^2} \mathbf{I}_{n_s} + \bar{\mathbf{C}} \mathbf{R}_{\bar{\mathbf{V}}}^{-1} \bar{\mathbf{C}}^H \right)^{-1} \right)}{n_s n_e} \\ &= \left(\frac{1}{\text{NMSE}_e^{(0)}} + \frac{P_1 T_1 / n_s}{(n_s - n_d) \sigma_{a,1}^2 \sigma_{h_e}^2 + \sigma_v^2} \right)^{-1}, \end{aligned} \tag{5.23}$$

where $\text{NMSE}_e^{(0)}$ was given in (5.9).

Notice, from (5.19) and (5.23), that the channel estimation performance at both the destination and the eavesdropper is affected by AN. However, this effect can be limited at the destination by improving its estimate in stage 0, i.e., by reducing $\text{NMSE}_d^{(0)}$, whereas that at the eavesdropper cannot. Yet, to decrease $\text{NMSE}_e^{(0)}$, the pilot signal power in stage 0 must be increased, which benefits the eavesdropper as well. Therefore the power allocation between pilot and AN signals in the two stages may have a significant impact on the DCE performance, and thus must be carefully determined.

Following [3], the optimal power allocation between pilot and AN signals can be determined by minimizing the NMSE at the destination subject to a lower constraint on the NMSE at the eavesdropper as well as a total power constraint. The problem can be formulated as follows:

$$\min_{P_0, P_1, \sigma_{a,1}^2} \text{NMSE}_d^{(1)} \tag{5.24a}$$

$$\text{subject to } \text{NMSE}_e^{(1)} \geq \gamma, \tag{5.24b}$$

$$P_0 T_0 + P_1 T_1 + \sigma_{a,1}^2 (n_s - n_d) T_1 \leq \bar{P}_{\text{ave}}(T_0 + T_1). \tag{5.24c}$$

where γ is a constraint on the smallest NMSE achievable at the eavesdropper and $\bar{P}_{\text{ave}}$ is the average power constraint.

Notice that the constraint in (5.24b) is interesting only when

$$\left(\frac{1}{\sigma_{h_e}^2} + \frac{P_{\text{ave}}(T_0 + T_1)}{n_s \sigma_v^2} \right)^{-1} \leq \gamma \leq \sigma_{h_e}^2. \tag{5.25}$$

The lower bound is given by the best NMSE achievable by the eavesdropper when all power is utilized to transmit pilot signals and no AN is used, and the upper bound is given by the worst NMSE achievable by the eavesdropper (i.e., the NMSE achieved when the mean of $\mathbf{H}_e$ is taken as its estimate). The constraint in (5.24b) would be inactive if γ is smaller than the former value and would be infeasible if it is greater than the latter value.

Let us define $\mathcal{E}_0 \triangleq P_0 T_0$ and $\mathcal{E}_1 \triangleq P_1 T_1$ as the energy of the pilot signals in stages 0 and 1, and define $\tilde{\gamma} \triangleq (\frac{1}{\gamma} - \frac{1}{\sigma_{h_e}^2}) n_s \sigma_v^2$ as an equivalent threshold constraint. In this case, the inequality in (5.25) can be written equivalently as

$$0 \leq \tilde{\gamma} \leq \bar{\mathcal{E}}_{\text{tot}} \triangleq P_{\text{ave}}(T_0 + T_1). \tag{5.26}$$

By substituting the above variables into (5.24) and by replacing $\text{NMSE}_d^{(1)}$ and $\text{NMSE}_e^{(1)}$ with their respective values in (5.19) and (5.23), the optimization problem can be written explicitly as

$$\max_{\mathcal{E}_0, \mathcal{E}_1, \sigma_{a,1}^2 \geq 0} \mathcal{E}_0 + \frac{(n_s \sigma_w^2 + \sigma_{h_d}^2 \mathcal{E}_0) \mathcal{E}_1}{n_s \sigma_w^2 + \sigma_{h_d}^2 \mathcal{E}_0 + n_s (n_s - n_d) \sigma_{h_d}^2 \sigma_{a,1}^2} \tag{5.27a}$$

$$\text{subject to } \mathcal{E}_0 + \frac{\sigma_v^2 \mathcal{E}_1}{(n_s - n_d) \sigma_{h_e}^2 \sigma_{a,1}^2 + \sigma_v^2} \leq \tilde{\gamma}, \tag{5.27b}$$

$$\mathcal{E}_0 + \mathcal{E}_1 + (n_s - n_d) \sigma_{a,1}^2 T_1 \leq \bar{\mathcal{E}}_{\text{tot}}. \tag{5.27c}$$

It has been shown in [3] that, when

$$\eta \triangleq n_s \left(\frac{\sigma_v^2}{\sigma_{h_e}^2} - \frac{\sigma_w^2}{\sigma_{h_d}^2} \right) > \tilde{\gamma},$$

the constraint in (5.27b) can be achieved with no need of AN. In this case, the optimal values are given by $(\sigma_{a,1}^2)^\star = 0$ and by $\mathcal{E}_0^\star$ and $\mathcal{E}_1^\star$ such that $\mathcal{E}_0^\star + \mathcal{E}_1^\star = \tilde{\gamma}$. When

$\eta \leq \tilde{\gamma}$, the problem can be converted to the one-variable optimization problem given below:

$$\max_{\mathcal{E}_0} \mathcal{E}_0 + \frac{(n_s \sigma_w^2 + \sigma_{h_d}^2 \mathcal{E}_0) \cdot \mathcal{E}_1(\mathcal{E}_0)}{n_s \sigma_w^2 + n_s(n_s - n_d)\sigma_{h_d}^2 \cdot \sigma_{a,1}^2(\mathcal{E}_0) + \sigma_{h_d}^2 \mathcal{E}_0} \tag{5.28a}$$

$$\text{subject to } \max\{\eta, 0\} \leq \mathcal{E}_0 \leq \tilde{\gamma}, \tag{5.28b}$$

where

$$\sigma_{a,1}^2(\mathcal{E}_0) = \frac{\bar{\mathcal{E}}_{\text{tot}} - \tilde{\gamma}}{(n_s - n_d)\left[T_1 + \sigma_{h_e}^2 (\tilde{\gamma} - \mathcal{E}_0)/\sigma_v^2\right]} \tag{5.29}$$

and

$$\mathcal{E}_1(\mathcal{E}_0) = \sigma_{h_e}^2 \left(\frac{\tilde{\gamma} - \mathcal{E}_0}{\sigma_v^2}\right)(n_s - n_d)\,\sigma_{a,1}^2(\mathcal{E}_0) + \tilde{\gamma} - \mathcal{E}_0. \tag{5.30}$$

The optimal value of $\mathcal{E}_0$ can be found via a one-dimensional line search over the finite interval $\max\{\eta, 0\} \leq \mathcal{E}_0 \leq \tilde{\gamma}$ and the optimal values of $\mathcal{E}_1$ and $\sigma_{a,1}^2$ are given by $\mathcal{E}_1(\mathcal{E}_0^\star)$ and $\sigma_{a,1}^2(\mathcal{E}_0^\star)$, respectively.

In the above, η can be viewed as a measure of channel quality difference between the main and the eavesdropper channels. When $\eta > \tilde{\gamma}$, the quality of the main channel is sufficiently higher than that of the eavesdropper channel, and thus the desired discrimination between the channel estimation performance at the two receivers can be naturally achieved (without the use of AN). When $\eta \leq \tilde{\gamma}$, the channel quality difference is relatively small, and thus AN would be required to achieve the constraint on the eavesdropper's NMSE. Detailed proofs can be found in [3].

Notice, from the optimization problem in (5.24) (and perhaps more clearly in (5.28)), that the power utilized for pilot transmission in stage 0, i.e., P_0 (or, equivalently, $\mathcal{E}_0$), is limited by the NMSE constraint at the eavesdropper. However, if P_0 is not sufficiently large, the preliminary channel estimate at the destination (which is fed back to the source) would not be sufficient for accurate AN placement. The NMSE achievable at the destination over the two-stage training would thus be limited as well. Interestingly, this can be easily overcome by repeating the feedback-and-retraining process multiple times to gradually refine the channel estimate at the destination (and thus the source) without violating the NMSE constraint at the eavesdropper.

Specifically, in the kth stage (where $k \geq 1$), the source emits similarly a training signal that consists of a pilot matrix and AN embedded in the null space of the estimated channel matrix. Given the channel estimate $\hat{\mathbf{H}}_{d,k-1}$ obtained over the previous $k - 1$ stages, the training signal can be expressed as

$$\mathbf{X}_k = \sqrt{\frac{P_k T_k}{n_s}} \mathbf{C}_k + \mathbf{N}_{\hat{\mathbf{H}}_{d,k-1}} \mathbf{A}_k, \tag{5.31}$$

where P_k is the pilot signal power, T_k is the training length in stage k, $\mathbf{C}_k \in \mathbb{C}^{n_s \times T_k}$ is the semi-unitary pilot matrix satisfying $\mathbf{C}_k \mathbf{C}_k^H = \mathbf{I}_{n_s}$, $\mathbf{N}_{\hat{\mathbf{H}}_{d,k-1}} \in \mathbb{C}^{n_s \times (n_s - n_d)}$ is a matrix whose columns form an orthonormal basis for the null space of $\hat{\mathbf{H}}_{d,k-1}$, and $\mathbf{A}_k \in \mathbb{C}^{(n_s - n_d) \times T_k}$ is an AN matrix with entries being i.i.d. $\mathcal{CN}(0, \sigma_{a,k}^2)$. The received signals at the destination and the eavesdropper are given by

$$\mathbf{Y}_k = \mathbf{H}_d \mathbf{X}_k + \mathbf{W}_k, \tag{5.32}$$

$$\mathbf{Z}_k = \mathbf{H}_e \mathbf{X}_k + \mathbf{V}_k, \tag{5.33}$$

where $\mathbf{W}_k \in \mathbb{C}^{n_d \times T_k}$ and $\mathbf{V}_k \in \mathbb{C}^{n_e \times T_k}$ are the AWGN matrices whose entries are i.i.d. Gaussian with 0 mean and variances σ_w^2 and σ_v^2, respectively.

Following the derivations in (5.19) and (5.23), the NMSE obtained by the destination and the eavesdropper after k stages of feedback-and-retraining can be expressed as

$$\text{NMSE}_d^{(k)} = \left(\frac{1}{\text{NMSE}_d^{(k-1)}} + \frac{P_k T_k / n_s}{\text{NMSE}_d^{(k-1)} (n_s - n_d)\sigma_{a,k}^2 + \sigma_w^2} \right)^{-1} \tag{5.34}$$

and

$$\text{NMSE}_e^{(k)} = \left(\frac{1}{\text{NMSE}_e^{(k-1)}} + \frac{P_k T_k / n_s}{(n_s - n_d)\sigma_{a,k}^2 \sigma_{h_e}^2 + \sigma_v^2} \right)^{-1} \tag{5.35}$$

$$= \left(\frac{1}{\sigma_{h_e}^2} + \frac{P_0 T_0}{n_s \sigma_v^2} + \sum_{\ell=1}^{k} \frac{P_\ell T_\ell / n_s}{(n_s - n_d)\sigma_{a,\ell}^2 \sigma_{h_e}^2 + \sigma_v^2} \right)^{-1}. \tag{5.36}$$

Notice, from (5.34), that the effect of AN at the destination in stage k can be reduced by decreasing the NMSE obtained in the previous $k - 1$ stages. This property is not enjoyed by the eavesdropper as can be seen in (5.36).

Suppose that K stages of feedback-and-retraining are employed. In this case, the optimal pilot signal and AN powers (i.e., $\{P_k\}_{k=0}^K$ and $\{\sigma_{a,k}^2\}_{k=1}^K$) can also be determined by minimizing the NMSE at the destination after K feedback-and-retraining stages, i.e., $\text{NMSE}_d^{(K)}$, subject to a lower constraint γ on the NMSE at the eavesdropper, i.e., $\text{NMSE}_e^{(K)}$, and a total power constraint. The optimization problem can be formulated as follows:

$$\min_{\{P_k\}_{k=0}^K, \{\sigma_{a,k}^2\}_{k=1}^K} \text{NMSE}_d^{(K)} \tag{5.37a}$$

$$\text{subject to } \text{NMSE}_e^{(K)} \geq \gamma, \tag{5.37b}$$

$$P_0 T_0 + \sum_{k=1}^{K} \left(P_k T_k + \sigma_{a,k}^2 (n_s - n_d) T_k \right) \leq \bar{P}_{ave} T_{tot} \tag{5.37c}$$

where $T_{tot} \triangleq \sum_{k=0}^{K} T_k$ is the total training length, $\bar{P}_{ave}$ is the average power constraint, and γ is the constraint on the smallest NMSE achievable at the eavesdropper. The optimization problem is nonconvex and is difficult to solve exactly. However, an approximate solution can be obtained efficiently by using the monomial approximation and the condensation method (i.e., a conversion of the problem into a series of geometric programming problems using successive convex approximations) [7]. Details can be found in [3].

Example 5.1 Let us consider an example in [3], where the source, the destination, and the eavesdropper are equipped with $n_s = 4$, n_d, and $n_e = 2$ antennas, respectively. The channel matrices $\mathbf{H}_d$ and $\mathbf{H}_e$ have entries that are i.i.d. Gaussian with zero mean and unit variance, i.e., $\sigma_{h_d}^2 = \sigma_{h_e}^2 = 1$. The AWGN matrices $\mathbf{W}_0, \mathbf{W}_1, \ldots, \mathbf{W}_K$ are also assumed to have entries that are i.i.d. Gaussian with zero mean and unit variance, i.e., $\sigma_w^2 = 1$. Here, the training length is given by $T_0 = T_1 = \cdots = T_K = \frac{300}{K+1}$, where K is the number of feedback-and-retraining stages.

In Fig. 5.2a, the NMSE at the destination and the eavesdropper is shown versus the total power constraint $\bar{P}_{ave}$ for different constraints at the eavesdropper, namely, $\gamma = 0.1$ and $\gamma = 0.03$. The number of feedback and retraining stages is $K = 11$. One can see that the NMSE at the destination decreases as the average transmit power increases while the eavesdropper's NMSE is bounded above by γ regardless of the value of $\bar{P}_{ave}$. In Fig. 5.2b, the optimal pilot and AN powers for the case with $K = 11$ feedback-and-retraining stages are shown for each stage of the process. We can see that both the pilot signal and AN powers gradually increase with the stage index. This is because, through the process, the source is able to gradually refine its knowledge of the channel, and thus AN can be placed in a gradually more accurate subspace to mask the pilot transmission. Hence, the source is more confident in expending resources for training in later stages.

5.2 Two-Way Training Discriminatory Channel Estimation Scheme

In the scheme described in Sect. 5.1, multiple feedback-and-retraining stages are needed to gradually refine source's knowledge of the channel. The number of stages required to achieve good performance increases as the constraint on the eavesdropper becomes more strict, and thus results in large training overhead, especially when the channel is fast fading. Interestingly, this inefficiency can be overcome by employing a two-way training based DCE scheme, where the destination also participates in the pilot signal transmission. Specifically, in the two-way training DCE scheme [4], the preliminary training signal is sent by the destination to enable channel estimation directly at the source. If the channel is reciprocal [8], the source will be able to infer knowledge of its forward channel to the destination from its estimate of the backward

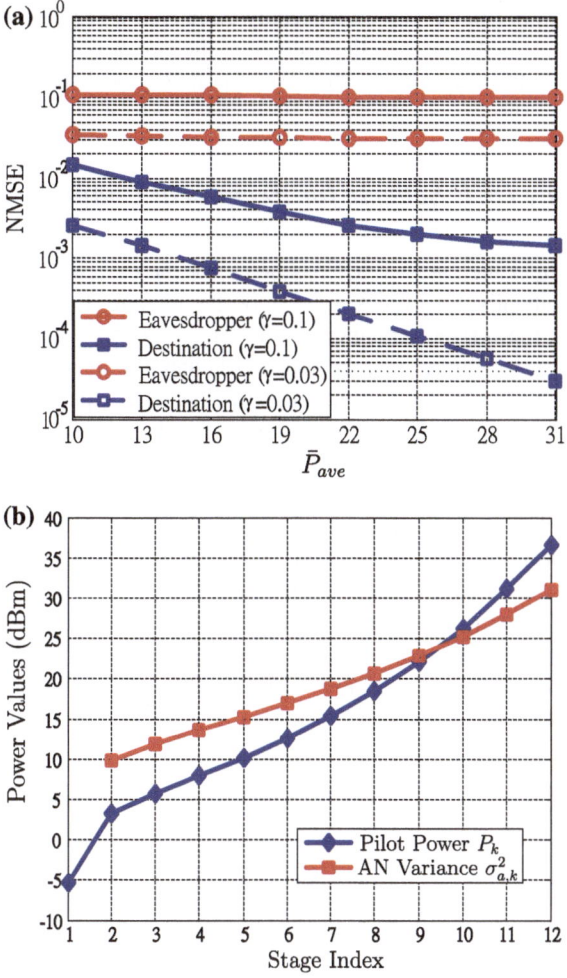

Fig. 5.2 NMSE performance and corresponding power allocation in the feedback-and-retraining DCE scheme with $K = 11$. **a** NMSE performance, **b** power allocation between pilot signals and AN for the case with $K = 11$

channel (i.e., the channel from the destination to the source). The AN-assisted training signal can then be transmitted in the next stage and the desired performance discrimination can be achieved in just two stages. If the channel is nonreciprocal, only an additional round-trip training stage is required on top of the original two-stage procedure to achieve the desired performance discrimination. The two-way DCE scheme is able to overcome the inefficiency in the previous scheme since, by having the destination emit the preliminary training signal instead of the source, the eavesdropper would not be able to obtain an estimate of the source-to-eavesdropper channel in this stage. Two-way training has been studied for conventional nonsecrecy applications, e.g., in [9], but can be especially effective for secrecy applications due to the reason described above. In the following, the two-way training DCE scheme is described for both cases with reciprocal and nonreciprocal channels.

5.2.1 Two-Way DCE Scheme for Reciprocal Channels

Let us first consider the two-way DCE scheme for the case where all channels are reciprocal. Specifically, let $\mathbf{H}_{sd} \in \mathbb{C}^{n_d \times n_s}$ be the channel from the source to the destination (i.e., the forward channel) and let $\mathbf{H}_{ds} \in \mathbb{C}^{n_s \times n_d}$ be the channel from the destination to the source (i.e., the backward channel), as illustrated in Fig. 5.3. When the channels are reciprocal, the forward channel is equal to the transpose of the backward channel, i.e., $\mathbf{H}_d \triangleq \mathbf{H}_{sd} = \mathbf{H}_{ds}^T$, and thus the source will be able to directly infer knowledge of the forward channel if it is able to obtain an estimate of the backward channel. Therefore, when the channels are reciprocal, the two-way DCE scheme can be performed using only two stages, namely, a backward training stage, where a preliminary training signal is sent from the destination to the source, and a forward training stage, where a pilot signal masked by AN is emitted by the source to achieve the discriminative channel estimation performance (similar to the retraining procedure in the feedback-and-retraining scheme).

Specifically, in the reverse training stage (i.e., stage 0), the destination emits an $n_d \times T_0$ training signal given by

$$\mathbf{X}_0 = \sqrt{\frac{P_0 T_0}{n_d}} \mathbf{C}_0, \tag{5.38}$$

where P_0 is the pilot signal power, T_0 is the reverse training length, and $\mathbf{C}_0 \in \mathbb{C}^{n_d \times T_0}$ is the semi-unitary pilot matrix that satisfies $\mathbf{C}_0 \mathbf{C}_0^H = \mathbf{I}_{n_d}$. The received signal at the source is

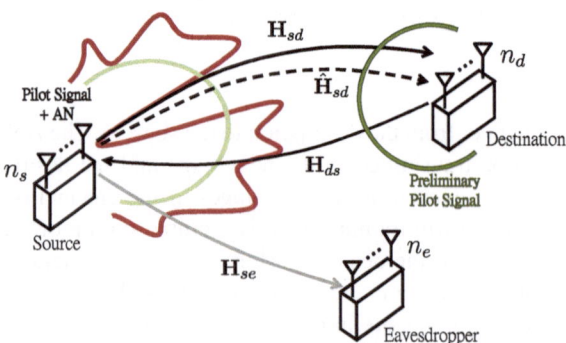

Fig. 5.3 Illustration of two-way training-based DCE schemes

$$\widetilde{\mathbf{Y}}_0 = \mathbf{H}_d^T \mathbf{X}_0 + \mathbf{U}_0, \tag{5.39}$$

where $\mathbf{U}_0 \in \mathbb{C}^{n_s \times T_0}$ is the AWGN at the source with entries being i.i.d. $\mathcal{CN}(0, \sigma_u^2)$ and $\mathbf{H}_d^T$ is the channel from the destination to the source whose entries are assumed to be i.i.d. with zero mean and variance $\sigma_{h_d}^2$. The reverse training signal sent by the destination allows the source to obtain an estimate of the forward channel by taking the transpose of its estimate of the backward channel. Therefore the LMMSE estimate of $\mathbf{H}_d$ at the source is given by

$$\hat{\mathbf{H}}_{d,0} = \left[\sigma_{h_d}^2 \widetilde{\mathbf{Y}}_0 (\sigma_{h_d}^2 \mathbf{X}_0^H \mathbf{X}_0 + \sigma_u^2 \mathbf{I}_{T_0})^{-1} \mathbf{X}_0^H \right]^T. \tag{5.40}$$

The estimation error matrix is defined as $\Delta\mathbf{H}_{d,0} = \hat{\mathbf{H}}_{d,0} - \mathbf{H}_d$ and its correlation is given by

$$E[\Delta\mathbf{H}_{d,0}^H \Delta\mathbf{H}_{d,0}] = n_d \left(\frac{1}{\sigma_{h_d}^2} + \frac{P_0 T_0}{n_d \sigma_u^2} \right)^{-1} \mathbf{I}_{n_s}. \tag{5.41}$$

The corresponding NMSE at the source in stage 0 is thus given by

$$\mathrm{NMSE}_d^{(0)} = \left(\frac{1}{\sigma_{h_d}^2} + \frac{P_0 T_0}{n_d \sigma_u^2} \right)^{-1}, \tag{5.42}$$

similar to that in (5.7).

Then, in the forward training stage, the source transmits a new training signal with AN embedded in the null space of the channel estimate $\hat{\mathbf{H}}_{d,0}$ to interfere with the reception at the eavesdropper. Similar to that in (5.10) and (5.31), the forward training signal can be written as

$$\mathbf{X}_1 = \sqrt{\frac{P_1 T_1}{n_s}} \mathbf{C}_1 + \mathbf{N}_{\hat{\mathbf{H}}_{d,0}} \mathbf{A}_1, \tag{5.43}$$

where P_1 is the pilot signal power, T_1 is the forward training length, $\mathbf{C}_1 \in \mathbb{C}^{n_s \times T_1}$ is the semi-unitary pilot matrix, $\mathbf{N}_{\hat{\mathbf{H}}_{d,0}} \in \mathbb{C}^{n_s \times (n_s - n_d)}$ is a matrix whose columns form an orthonormal basis for the null space of $\hat{\mathbf{H}}_{d,0}$, and $\mathbf{A}_1 \in \mathbb{C}^{(n_s - n_d) \times T_1}$ is an AN matrix with entries being i.i.d. $\mathcal{CN}(0, \sigma_{a,1}^2)$. The received signals at the destination and the eavesdropper can be written as

$$\mathbf{Y}_1 = \mathbf{H}_d \mathbf{X}_1 + \mathbf{W}_1 = \sqrt{\frac{P_1 T_1}{n_s}} \mathbf{H}_d \mathbf{C}_1 + \mathbf{H}_d \mathbf{N}_{\hat{\mathbf{H}}_{d,0}} \mathbf{A}_1 + \mathbf{W}_1, \tag{5.44}$$

$$\mathbf{Z}_1 = \mathbf{H}_e \mathbf{X}_1 + \mathbf{V}_1 = \sqrt{\frac{P_1 T_1}{n_s}} \mathbf{H}_e \mathbf{C}_1 + \mathbf{H}_e \mathbf{N}_{\hat{\mathbf{H}}_{d,0}} \mathbf{A}_1 + \mathbf{V}_1, \tag{5.45}$$

where $\mathbf{W}_1 \in \mathbb{C}^{n_d \times T_1}$ and $\mathbf{V}_1 \in \mathbb{C}^{n_e \times T_1}$ are the AWGN matrices at the destination and the eavesdropper, respectively, with entries that are i.i.d. Gaussian with 0 mean and variances σ_w^2 and σ_v^2, respectively. The channel estimation at the destination and the eavesdropper is then computed based on their respective observations in stage 1, i.e., $\mathbf{Y}_1$ and $\mathbf{Z}_1$. Notice that this is different from the feedback-and-retraining DCE scheme where the received signals in both stages are utilized for channel estimation.

Specifically, let $\bar{\mathbf{C}} \triangleq \sqrt{\frac{P_1 T_1}{n_s}} \mathbf{C}_1$ be the equivalent pilot matrix and let $\bar{\mathbf{W}} \triangleq \mathbf{H}_d \mathbf{N}_{\hat{\mathbf{H}}_{d,0}} \mathbf{A}_1 + \mathbf{W}_1 = -\Delta \mathbf{H}_{d,0} \mathbf{N}_{\hat{\mathbf{H}}_{d,0}} \mathbf{A}_1 + \mathbf{W}_1$ and $\bar{\mathbf{V}} \triangleq \mathbf{H}_e \mathbf{N}_{\hat{\mathbf{H}}_{d,0}} \mathbf{A}_1 + \mathbf{V}_1$ be the equivalent noise at the destination and the eavesdropper, respectively. Then, the LMMSE estimate at the destination can be expressed as

$$\hat{\mathbf{H}}_{d,1} = n_d \sigma_{h_d}^2 \mathbf{Y}_1 \left(n_d \sigma_{h_d}^2 \bar{\mathbf{C}}^H \bar{\mathbf{C}} + \mathbf{R}_{\bar{\mathbf{W}}} \right)^{-1} \bar{\mathbf{C}}^H, \tag{5.46}$$

where

$$\mathbf{R}_{\bar{\mathbf{W}}} \triangleq \mathrm{E}\left[\bar{\mathbf{W}}^H \bar{\mathbf{W}} \right] = \left(\mathrm{E}[\|\Delta \mathbf{H}_{d,0} \mathbf{N}_{\hat{\mathbf{H}}_{d,0}}\|_F^2] \sigma_{a,1}^2 + n_d \sigma_w^2 \right) \mathbf{I}_{T_1} \tag{5.47}$$

$$= \left[n_d(n_s - n_d) \left(\frac{1}{\sigma_{h_d}^2} + \frac{P_0 T_0}{n_d \sigma_u^2} \right)^{-1} \sigma_{a,1}^2 + n_d \sigma_w^2 \right] \mathbf{I}_{T_1} \tag{5.48}$$

$$= \left[n_d(n_s - n_d) \mathrm{NMSE}_d^{(0)} \sigma_{a,1}^2 + n_d \sigma_w^2 \right] \mathbf{I}_{T_1} \tag{5.49}$$

is the correlation matrix of $\bar{\mathbf{W}}$. The associated NMSE is given by

$$\mathrm{NMSE}_d^{(1)} \triangleq \frac{\mathrm{tr}\left(\mathrm{E}\left[\Delta \mathbf{H}_{d,1}^H \Delta \mathbf{H}_{d,1} \right] \right)}{n_s n_d} = \frac{\mathrm{tr}\left(\left(\frac{1}{n_d \sigma_{h_d}^2} \mathbf{I}_{n_s} + \bar{\mathbf{C}} \mathbf{R}_{\bar{\mathbf{W}}}^{-1} \bar{\mathbf{C}}^H \right)^{-1} \right)}{n_s n_d}$$

$$= \left(\frac{1}{\sigma_{h_d}^2} + \frac{P_1 T_1 / n_s}{\mathrm{NMSE}_d^{(0)} \cdot (n_s - n_d) \sigma_{a,1}^2 + \sigma_w^2} \right)^{-1}, \tag{5.50}$$

Similarly, the LMMSE estimate at the eavesdropper is given by

$$\hat{\mathbf{H}}_{e,1} = n_e \sigma_{h_e}^2 \mathbf{Z}_1 \left(n_e \sigma_{h_e}^2 \bar{\mathbf{C}}^H \bar{\mathbf{C}} + \mathbf{R}_{\bar{\mathbf{V}}} \right)^{-1} \bar{\mathbf{C}}^H \tag{5.51}$$

where $\bar{\mathbf{V}} \triangleq \mathbf{H}_e \mathbf{N}_{\hat{\mathbf{H}}_{d,0}} \mathbf{A}_1 + \mathbf{V}_1$ is the equivalent noise matrix and $\mathbf{R}_{\bar{\mathbf{V}}} \triangleq \mathrm{E}[\bar{\mathbf{V}}^H \bar{\mathbf{V}}] = n_e \left[(n_s - n_d) \sigma_{a,1}^2 \sigma_{h_e}^2 + \sigma_v^2 \right] \mathbf{I}_{T_1}$ is the correlation matrix of $\bar{\mathbf{V}}$. The NMSE of the estimate at the eavesdropper is given by

$$\text{NMSE}_e^{(1)} = \left(\frac{1}{\sigma_{h_e}^2} + \frac{P_1 T_1 / n_s}{(n_s - n_d)\sigma_{a,1}^2 \sigma_{h_e}^2 + \sigma_v^2} \right)^{-1}. \tag{5.52}$$

It can be observed from (5.50) and (5.52) that training in stage 0 benefits the channel estimation at the destination by reducing the AN interference that it receives, but does not benefit that at the eavesdropper. Hence, when the source and the destination are limited only by individual power constraints, the destination will be able to expend all its resources for reverse training without the concern of benefiting the eavesdropper. However, when a total power constraint exists, increasing the pilot signal power in the reverse training stage reduces the power that can be used for pilot and AN transmissions in the forward training stage. Following the approach adopted in the previous section and in [4], the optimal power allocation can be found by minimizing the NMSE at the destination subject to a lower constraint on the NMSE at the eavesdropper. Let us define $\mathcal{E}_0 \triangleq P_0 T_0$ and $\mathcal{E}_1 \triangleq P_1 T_1$ as the pilot signal energy in stage 0 (at the destination) and in stage 1 (at the destination), respectively. The optimization problem can be formulated as follows:

$$\min_{\mathcal{E}_0, \mathcal{E}_1, \sigma_{a,1}^2 \geq 0} \text{NMSE}_d^{(1)} \tag{5.53a}$$

$$\text{subject to } \text{NMSE}_e^{(1)} \geq \gamma, \tag{5.53b}$$

$$\mathcal{E}_0 + \mathcal{E}_1 + (n_s - n_d)\sigma_{a,1}^2 T_1 \leq \bar{\mathcal{E}}_{tot}, \tag{5.53c}$$

$$\mathcal{E}_0 \leq \bar{\mathcal{E}}_d, \tag{5.53d}$$

$$\mathcal{E}_1 + (n_s - n_d)\sigma_{a,1}^2 T_1 \leq \bar{\mathcal{E}}_s, \tag{5.53e}$$

where γ is the lower constraint on the NMSE at the eavesdropper, $\bar{\mathcal{E}}_{tot}$ is the total energy constraint, and $\bar{\mathcal{E}}_d$ and $\bar{\mathcal{E}}_s$ are the individual energy constraints at the destination and the source, respectively. Similar to that in the feedback-and-retraining DCE scheme, the NMSE constraint γ should be chosen such that

$$\left(\frac{1}{\sigma_{h_e}^2} + \frac{\min\{\bar{\mathcal{E}}_s, \bar{\mathcal{E}}_{tot}\}}{n_s \sigma_v^2} \right)^{-1} \leq \gamma \leq \sigma_{h_e}^2, \tag{5.54}$$

where the value on the left-hand-side is the best NMSE achievable by the eavesdropper when all power is utilized by the source to transmit pilot signals and that on the right-hand-side is given by the worst NMSE achievable by the eavesdropper. By defining $\tilde{\gamma} \triangleq (\frac{1}{\gamma} - \frac{1}{\sigma_{h_e}^2})n_s \sigma_v^2$, the condition in (5.54) can be expressed equivalently as

$$0 \leq \tilde{\gamma} \leq \min\{\bar{\mathcal{E}}_s, \bar{\mathcal{E}}_{tot}\} \tag{5.55}$$

and the eavesdropper's NMSE constraint in (5.53b) can be written as

$$\frac{\sigma_v^2 \mathcal{E}_1}{(n_s - n_d)\sigma_{h_e}^2 \sigma_{a,1}^2 + \sigma_v^2} \leq \tilde{\gamma} \tag{5.56}$$

or, equivalently, as

$$\mathcal{E}_1 \leq \tilde{\gamma}\left[(n_s - n_d)\sigma_{h_e}^2 \sigma_{a,1}^2/\sigma_v^2 + 1\right]. \tag{5.57}$$

Let us first consider the case where only individual power constraints exist (or when the total power constraint is redundant). It has been shown in [4] that, when the difference between the main and the eavesdropper channel quality is large enough such that

$$\eta \triangleq n_d\left(\frac{\sigma_v^2}{\sigma_{h_e}^2}\frac{\sigma_u^2}{\sigma_w^2} - \frac{\sigma_u^2}{\sigma_{h_d}^2}\right) > \bar{\mathcal{E}}_d, \tag{5.58}$$

the NMSE at the destination and the constraint at the eavesdropper can be satisfied without the use of AN (and thus without the need for reverse training). In this case, it is sufficient to set the optimal reverse training energy and the optimal AN power in the forward training stage to 0, i.e., $\mathcal{E}_0^\star = 0$ and $(\sigma_{a,1}^2)^\star = 0$. The optimal pilot signal energy utilized in the forward training is then given by $\mathcal{E}_1^\star = \tilde{\gamma}$. On the other hand, when $\eta \leq \bar{\mathcal{E}}_d$, the use of AN can help achieve the desired performance discrimination and its effectiveness is maximized by utilizing all the available energy at the destination for reverse training, i.e., the optimal reverse training energy is $\mathcal{E}_0^\star = \bar{\mathcal{E}}_d$. In this case, the optimal pilot signal energy and AN power in the forward training should be given by [4]

$$\mathcal{E}_1^\star = \bar{\mathcal{E}}_s - \frac{(\bar{\mathcal{E}}_s - \tilde{\gamma})T_1}{T_1 + \tilde{\gamma}\sigma_{h_e}^2/\sigma_v^2} \tag{5.59}$$

and

$$(\sigma_{a,1}^2)^\star = \frac{\bar{\mathcal{E}}_s - \tilde{\gamma}}{(T_1 + \tilde{\gamma}\sigma_{h_e}^2/\sigma_v^2)(n_s - n_d)}. \tag{5.60}$$

Next, let us consider the case where the total power constraint is of interest as well (i.e., the case where $\max\{\bar{\mathcal{E}}_s, \bar{\mathcal{E}}_d\} \leq \bar{\mathcal{E}}_{tot} \leq \bar{\mathcal{E}}_s + \bar{\mathcal{E}}_d$). It has also been shown in [4] that, when the channel quality difference is large enough such that $\eta > \min\{\bar{\mathcal{E}}_d, \bar{\mathcal{E}}_{tot} - \tilde{\gamma}\}$, the use of AN is not necessary and the optimal power values are given by $\mathcal{E}_0^\star = 0$, $(\sigma_{a,1}^2)^\star = 0$, and $\mathcal{E}_1^\star = \tilde{\gamma}$, similar to that in the previous case. On the other hand, when $\eta \leq \min\{\bar{\mathcal{E}}_d, \bar{\mathcal{E}}_{tot} - \tilde{\gamma}\}$, the problem can be converted to a one-dimensional optimization problem, similar to that in (5.28). The one-dimensional optimization problem is given as follows:

$$\max_{\mathcal{E}_0} \frac{(n_d \sigma_u^2 + \sigma_{h_d}^2 \mathcal{E}_0) \cdot \mathcal{E}_1(\mathcal{E}_0)}{n_d \sigma_u^2 + \sigma_{h_d}^2 \mathcal{E}_0 + n_d(n_s - n_d)\sigma_{h_d}^2 \frac{\sigma_u^2}{\sigma_w^2} \cdot \sigma_{a,1}^2(\mathcal{E}_0)} \tag{5.61a}$$

$$\text{subject to } \max\{0, \eta, \bar{\mathcal{E}}_{tot} - \bar{\mathcal{E}}_s\} \le \mathcal{E}_0 \le \min\{\bar{\mathcal{E}}_d, \bar{\mathcal{E}}_{tot} - \tilde{\gamma}\}, \tag{5.61b}$$

where

$$\mathcal{E}_1(\mathcal{E}_0) \triangleq \tilde{\gamma}\left(\frac{\sigma_{h_e}^2}{\sigma_v^2} \frac{\bar{\mathcal{E}}_{tot} - \tilde{\gamma} - \mathcal{E}_0}{T_1 + \sigma_{h_e}^2 \tilde{\gamma}/\sigma_v^2} + 1\right) \tag{5.62}$$

and

$$\sigma_{a,1}^2(\mathcal{E}_0) \triangleq \frac{1}{n_s - n_d} \cdot \frac{\bar{\mathcal{E}}_{tot} - \tilde{\gamma} - \mathcal{E}_0}{T_1 + \sigma_{h_e}^2 \tilde{\gamma}/\sigma_v^2}. \tag{5.63}$$

The optimal reverse training energy, i.e., $\mathcal{E}_0^\star$, can be obtained by performing a line search over the finite interval given in (5.61b). The optimal pilot energy and AN power in the forward training are then given by $\mathcal{E}_1^\star(\mathcal{E}_0^\star)$ and $(\sigma_{a,1}^2)^\star(\mathcal{E}_0^\star)$.

Example 5.2 Let us consider an example, given in [4], where the source, the destination, and the eavesdropper are equipped with $n_s = 4$, $n_d = 2$, and $n_e = 2$ antennas, respectively. The channel matrices $\mathbf{H}_d$ and $\mathbf{H}_e$ are all assumed to have entries that are i.i.d. Gaussian with zero mean and unit variance, i.e., $\sigma_{h_d}^2 = \sigma_{h_e}^2 = 1$. The AWGN matrices are also assumed to have entries that are i.i.d. Gaussian with zero mean and unit variance, i.e., $\sigma_u^2 = \sigma_w^2 = \sigma_v^2 = 1$. The training lengths are chosen as $T_0 = n_d = 2$ and $T_1 = n_s = 4$, and therefore the total lengths of the training signals emitted by the source and the destination are also given by $T_{s,tot} = T_1 = 4$ and $T_{d,tot} = T_0 = 2$, respectively. Let $\bar{P}_{ave} \triangleq \bar{\mathcal{E}}_{tot}/(T_{s,tot} + T_{d,tot})$ be the overall average transmit power constraint. Moreover, let $\bar{P}_s \triangleq \bar{\mathcal{E}}_s/T_{s,tot} = 30\text{dB}$ and $\bar{P}_d \triangleq \bar{\mathcal{E}}_d/T_{d,tot} = 20\text{dB}$ (relative to the noise variance) be the average power constraints at the source and the destination, respectively.

In Fig. 5.4a, the NMSEs at the destination and the eavesdropper are shown for varying values of the overall average power constraint $\bar{P}_{ave}$. The curves are shown for two different NMSE constraints at the eavesdropper, namely, the case with $\gamma = 0.1$ and the case with $\gamma = 0.03$. One can observe that the NMSE at the destination improves as $\bar{P}_{ave}$ increases while the NMSE at the eavesdropper is successfully constrained above the value γ. However, the NMSE at the destination eventually saturates due to constraints on the individual powers at the source and the destination. In Fig. 5.4b, the corresponding pilot signal power values, i.e., P_0 and P_1, and the AN power $(n_s - n_d)\sigma_{a,1}^2$ are shown for varying values of $\bar{P}_{ave}$. We can see that, as the NMSE constraint at the eavesdropper becomes more strict, the AN power should be increased, and thus the power allocated to obtaining preliminary training (i.e., the power utilized in stage 0) should also increase, so that a better preliminary channel estimate can be obtained.

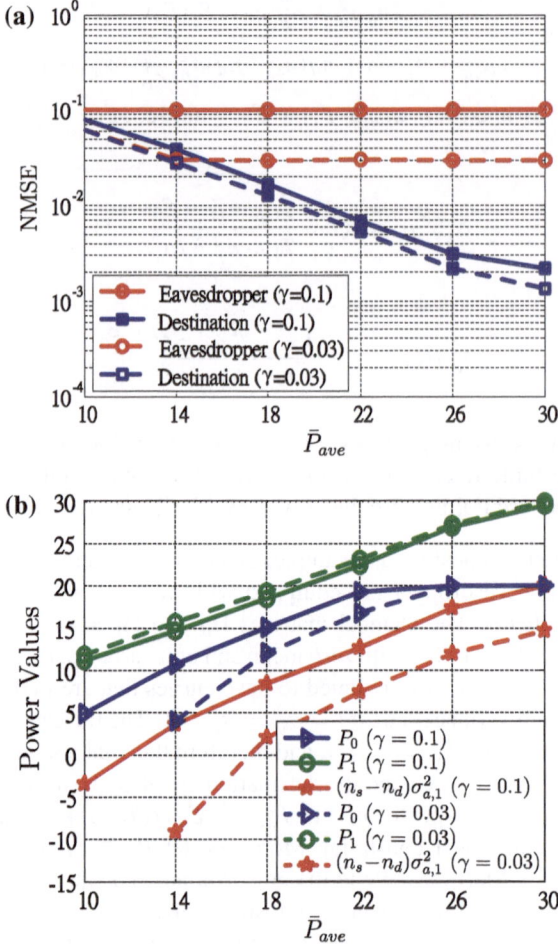

Fig. 5.4 NMSE performance and corresponding power allocation in the two-way DCE scheme for reciprocal channels. **a** NMSE performance, **b** power allocation between the pilot signal and AN powers

5.2.2 Two-Way DCE Scheme for Nonreciprocal Channels

When the channel is nonreciprocal, the source cannot directly infer knowledge of the forward channel with its estimate of the backward channel in the reverse training stage. In this case, an additional round-trip training must be employed. In particular, in the round-trip training stage, a random pilot matrix is emitted by the source and then echoed back by the destination using an amplify-and-forward strategy. With the echoed signal received from the destination, the source will be able to obtain an estimate of the combined backward and forward channels, i.e., $\mathbf{H}_{ds}\mathbf{H}_{sd}$.

Then, together with the estimate of the backward channel $\mathbf{H}_{ds}$ obtained in the reverse training stage, an estimate of the forward channel matrix $\mathbf{H}_{sd}$ can be computed. Consequently, the two-way DCE scheme in the case of nonreciprocal channels consists of three stages: a backward training stage, a round-trip training stage, and a forward training stage. The backward and forward training stages are similar to that in case with reciprocal channels.

Specifically, in the backward training stage (i.e., stage 0), the destination emits training signal

$$\mathbf{X}_0 = \sqrt{\frac{P_0 T_0}{n_d}} \mathbf{C}_0, \qquad (5.64)$$

where P_0 is the pilot signal power, T_0 is the reverse training length, and $\mathbf{C}_0 \in \mathbb{C}^{n_d \times T_0}$ is the semi-unitary pilot matrix that satisfies $\mathbf{C}_0 \mathbf{C}_0^H = \mathbf{I}_{n_d}$. The received signal at the source is

$$\widetilde{\mathbf{Y}}_0 = \mathbf{H}_{ds} \mathbf{X}_0 + \mathbf{U}_0, \qquad (5.65)$$

where $\mathbf{U}_0 \in \mathbb{C}^{n_s \times T_0}$ is the AWGN matrix with entries that are i.i.d. $\mathcal{CN}(0, \sigma_u^2)$ and $\mathbf{H}_{ds}$ is the channel from the destination to the source whose entries are assumed to be i.i.d. with zero mean and variance $\sigma_{h_{ds}}^2$. Based on the received signal $\widehat{\mathbf{Y}}_0$, the source can obtain an estimate of the backward channel matrix $\mathbf{H}_{ds}$. In particular, the LMMSE estimate of $\mathbf{H}_{ds}$ at the source is given by

$$\hat{\mathbf{H}}_{ds,0} = \sigma_{h_{ds}}^2 \widetilde{\mathbf{Y}}_0 \left(\sigma_{h_{ds}}^2 \mathbf{X}_0^H \mathbf{X}_0 + \sigma_u^2 \mathbf{I}_{T_0} \right)^{-1} \mathbf{X}_0^H. \qquad (5.66)$$

The estimation error matrix is defined as $\Delta \mathbf{H}_{ds,0} = \hat{\mathbf{H}}_{ds,0} - \mathbf{H}_{ds}$ and its correlation is given by

$$\mathrm{E}[\Delta \mathbf{H}_{ds,0}^H \Delta \mathbf{H}_{ds,0}] = n_s \left(\frac{1}{\sigma_{h_{ds}}^2} + \frac{P_0 T_0}{n_d \sigma_u^2} \right)^{-1} \mathbf{I}_{n_d}. \qquad (5.67)$$

The NMSE of the estimate at the source in stage 0 is thus given by

$$\mathrm{NMSE}_{ds}^{(0)} = \frac{\mathrm{tr}\left(\mathrm{E}[\Delta \mathbf{H}_{ds,0}^H \Delta \mathbf{H}_{ds,0}] \right)}{n_s n_d} = \left(\frac{1}{\sigma_{h_{ds}}^2} + \frac{P_0 T_0}{n_d \sigma_u^2} \right)^{-1}, \qquad (5.68)$$

similar to that in (5.42). Notice that, when the channel is nonreciprocal, the source will not be able to infer knowledge of the forward channel $\mathbf{H}_{sd}$ from the estimate of $\mathbf{H}_{ds}$. Hence, an additional round-trip training stage is required to provide the source with information of the forward channel.

In the round-trip training stage (i.e., stage 1), the source first emits a random training signal that is generated upon transmission and is known only to itself. Then, this signal is echoed back by the destination using an amplify-and-forward strategy. The training signal emitted by the source in stage 1 is given by

$$\mathbf{X}_{s1} = \sqrt{\frac{P_{s1}T_1}{n_s}}\mathbf{C}_1, \tag{5.69}$$

P_{s1} is the transmit power at the source in stage 1 and T_1 is the training length. For convenience, we choose $T_1 = n_s$ and choose $\mathbf{C}_1 \in \mathbb{C}^{n_s \times n_s}$ as a randomly generated unitary pilot matrix with $\mathbf{C}_1\mathbf{C}_1^H = \mathbf{C}_1^H\mathbf{C}_1 = \mathbf{I}_{n_s}$. The received signal at the destination is given by

$$\mathbf{Y}_1 = \mathbf{H}_{sd}\mathbf{X}_{s1} + \mathbf{W}_1, \tag{5.70}$$

where $\mathbf{W}_1 \in \mathbb{C}^{n_d \times T_1}$ is the AWGN matrix at the destination with entries that are i.i.d. $\mathcal{CN}(0, \sigma_w^2)$,. Upon receiving $\mathbf{Y}_1$, the destination forwards an amplified version of the signal to the source. The echoed signal is given by

$$\mathbf{X}_{d1} = \alpha\mathbf{Y}_1, \tag{5.71}$$

where

$$\alpha = \sqrt{\frac{P_{d1}T_1}{P_{s1}T_1 n_d \sigma_{h_{sd}}^2 + T_1 n_d \sigma_w^2}} \tag{5.72}$$

is the amplifying gain and P_{d1} is the transmit power at the destination in stage 1. The received signal at the source can be written as

$$\widetilde{\mathbf{Y}}_1 = \mathbf{H}_{ds}\mathbf{X}_{d1} + \mathbf{U}_1 \tag{5.73}$$

$$= \alpha\mathbf{H}_{ds}\mathbf{H}_{sd}\mathbf{X}_{s1} + \alpha\mathbf{H}_{ds}\mathbf{W}_1 + \mathbf{U}_1 \tag{5.74}$$

$$= \alpha(\hat{\mathbf{H}}_{ds,0} - \Delta\mathbf{H}_{ds,0})\mathbf{H}_{sd}\mathbf{X}_{s1} + \alpha(\hat{\mathbf{H}}_{ds,0} - \Delta\mathbf{H}_{ds,0})\mathbf{W}_1 + \mathbf{U}_1 \tag{5.75}$$

where $\mathbf{U}_1 \in \mathbb{C}^{n_s \times T_1}$ is the AWGN matrix at the source in stage 1 with entries that are i.i.d. $\mathcal{CN}(0, \sigma_u^2)$, and $\Delta\mathbf{H}_{ds,0} \triangleq \hat{\mathbf{H}}_{ds,0} - \mathbf{H}_{ds}$ is the error of the estimate in stage 0. The estimation error and the AWGN terms can be collected into an effective noise term

$$\bar{\mathbf{U}}_1 \triangleq \alpha\hat{\mathbf{H}}_{ds,0}\mathbf{W}_1 - \Delta\mathbf{H}_{ds,0}\mathbf{H}_{sd}\mathbf{X}_{s1} - \alpha\Delta\mathbf{H}_{ds,0}\mathbf{W}_1 + \mathbf{U}_1, \tag{5.76}$$

and thus the received signal at the source can be written equivalently as

$$\widetilde{\mathbf{Y}}_1 = \alpha\hat{\mathbf{H}}_{ds,0}\mathbf{H}_{sd}\mathbf{X}_{s1} + \bar{\mathbf{U}}_1. \tag{5.77}$$

Notice that, due to the multiplication of $\mathbf{H}_{ds}$, the received signal on each antenna at the source contains information of all entries in the forward channel matrix, i.e., $\mathbf{H}_{sd}$. Therefore, it is necessary to consider all entries of the received signal $\widetilde{\mathbf{Y}}_1$ when computing the estimate of each entry of $\mathbf{H}_{sd}$.

Let us define $\tilde{\mathbf{y}}_1 \triangleq \mathrm{vec}(\widetilde{\mathbf{Y}}_1)$, $\mathbf{h}_{sd} \triangleq \mathrm{vec}(\mathbf{H}_{sd})$, $\mathbf{w}_1 = \mathrm{vec}(\mathbf{W}_1)$, and $\mathbf{u}_1 = \mathrm{vec}(\mathbf{U}_1)$ as the vectorized counterparts of $\widetilde{\mathbf{Y}}_1$, $\mathbf{H}_{sd}$, $\mathbf{W}_1$, and $\mathbf{U}_1$ obtained by stacking the column vectors of each corresponding matrix. By the fact that $\mathrm{vec}(\mathbf{ABC}) = (\mathbf{C}^T \otimes \mathbf{A})$

vec($\mathbf{B}$), where $\otimes$ represents the Kronecker product [10], the received signal in (5.77) can be rewritten as

$$\tilde{\mathbf{y}}_{t1} = \alpha(\mathbf{X}_{s1}^T \otimes \hat{\mathbf{H}}_{ds,0})\mathbf{h}_{sd} + \alpha(\mathbf{I}_{T_1} \otimes \hat{\mathbf{H}}_{ds,0})\mathbf{w}_1$$
$$- \alpha(\mathbf{X}_{s1}^T \otimes \Delta\mathbf{H}_{ds,0})\mathbf{h}_{sd} - \alpha(\mathbf{I}_{T_1} \otimes \Delta\mathbf{H}_{ds,0})\mathbf{w}_1 + \mathbf{u}_1. \tag{5.78}$$

Then, given $\hat{\mathbf{H}}_{ds,0}$, the LMMSE estimate of $\mathbf{H}_{sd}$ in stage 1 can be computed as [4]

$$\hat{\mathbf{h}}_{sd,1} = \frac{1}{\alpha\sigma_w^2}\left(\frac{1}{\sigma_{h_{sd}}^2} + \frac{P_{s1}T_1}{n_s\sigma_w^2}\right)^{-1}\left[\mathbf{X}_{s1}^* \otimes \hat{\mathbf{H}}_{ds,0}^H\left(\hat{\mathbf{H}}_{ds,0}\hat{\mathbf{H}}_{ds,0}^H + \beta\mathbf{I}_{n_s}\right)^{-1}\right]\tilde{\mathbf{y}}_1, \tag{5.79}$$

where

$$\beta \triangleq n_d\left(\frac{1}{\sigma_{h_{ds}}^2} + \frac{P_0T_0}{n_d\sigma_u^2}\right)^{-1} + \frac{\sigma_u^2}{\alpha^2\sigma_{h_{sd}}^2\sigma_w^2}\left(\frac{1}{\sigma_{h_{sd}}^2} + \frac{P_{s1}T_1}{n_s\sigma_w^2}\right)^{-1}. \tag{5.80}$$

The estimate can be written in the original matrix form as

$$\hat{\mathbf{H}}_{sd,1} = \frac{1}{\alpha\sigma_w^2}\left(\frac{1}{\sigma_{h_{sd}}^2} + \frac{P_{s1}T_1}{n_s\sigma_w^2}\right)^{-1}\hat{\mathbf{H}}_{ds,0}^H\left(\hat{\mathbf{H}}_{ds,0}\hat{\mathbf{H}}_{ds,0}^H + \beta\mathbf{I}_{n_s}\right)^{-1}\tilde{\mathbf{Y}}_1\mathbf{X}_{s1}^H. \tag{5.81}$$

The correlation of $\Delta\mathbf{h}_{sd,1}$ conditioned on $\hat{\mathbf{H}}_{ds,0}$ is given by

$$E\left[\Delta\mathbf{h}_{sd,1}\Delta\mathbf{h}_{sd,1}^H\Big|\hat{\mathbf{H}}_{ds,0}\right]$$
$$= \mathbf{I}_{n_s} \otimes \left\{\sigma_{h_{sd}}^2\mathbf{I}_{n_d} - \frac{\sigma_{h_{sd}}^4 P_{s1}T_1}{\sigma_{h_{sd}}^2 P_{s1}T_1 + n_s\sigma_w^2}\left[(\frac{1}{\beta}\hat{\mathbf{H}}_{ds,0}^H\hat{\mathbf{H}}_{ds,0})^{-1} + \mathbf{I}_{n_d}\right]^{-1}\right\}. \tag{5.82}$$

Finally, in the forward training stage (i.e., stage 2), the source emits a new training signal with AN embedded in the null space of the estimated channel matrix $\hat{\mathbf{H}}_{sd,1}$. Similar to (5.10) and (5.43), the training signal emitted by the source can be written as

$$\mathbf{X}_2 = \sqrt{\frac{P_2T_2}{n_s}}\mathbf{C}_2 + \mathbf{N}_{\hat{\mathbf{H}}_{sd,1}}\mathbf{A}_2, \tag{5.83}$$

where P_2 is the pilot signal power, T_2 is the forward training length, $\mathbf{C}_2 \in \mathbb{C}^{n_s \times T_1}$ is the semi-unitary pilot matrix chosen such that $\mathbf{C}_2\mathbf{C}_2^H = \mathbf{I}_{n_s}$, $\mathbf{N}_{\hat{\mathbf{H}}_{sd,1}} \in \mathbb{C}^{n_s \times (n_s-n_d)}$ is a matrix whose columns form an orthonormal basis for the null space of $\hat{\mathbf{H}}_{d,0}$, and $\mathbf{A}_2 \in \mathbb{C}^{(n_s-n_d) \times T_1}$ is an AN matrix with entries being i.i.d. $\mathcal{CN}(0, \sigma_{a,2}^2)$. Notice that, different from stage 1, the pilot matrix $\mathbf{C}_2$ emitted in stage 2 is assumed to be known at all terminals. The received signals at the destination and the eavesdropper can be written as

$$Y_2 = H_{sd}X_2 + W_2 = \sqrt{\frac{P_2 T_2}{n_s}} H_{sd}C_2 + H_{sd}N_{\hat{H}_{sd,1}} A_2 + W_2, \tag{5.84}$$

$$Z_2 = H_{se}X_2 + V_2 = \sqrt{\frac{P_2 T_2}{n_s}} H_{se}C_2 + H_{se}N_{\hat{H}_{sd,1}} A_2 + V_2, \tag{5.85}$$

where $W_2 \in \mathbb{C}^{n_d \times T_2}$ and $V_2 \in \mathbb{C}^{n_e \times T_2}$ are the AWGN matrices with entries that are i.i.d. with distributions $\mathcal{CN}(0, \sigma_w^2)$ and $\mathcal{CN}(0, \sigma_v^2)$, respectively.

To evaluate the LMMSE estimate at the destination, let us define similarly $y_2 \triangleq \mathrm{vec}(Y_2)$, $h_{sd} \triangleq \mathrm{vec}(H_{sd})$, $\Delta h_{sd,1} \triangleq \mathrm{vec}(\Delta H_{sd,1})$, and $w_2 = \mathrm{vec}(W_2)$ as the vectorized counterparts of their respective matrices. By letting $\bar{C}_2 \triangleq \sqrt{\frac{P_2 T_2}{n_s}} C_2$ and by the fact that $H_{sd}N_{\hat{H}_{sd,1}} A_2 = -\Delta H_{sd,1}N_{\hat{H}_{sd,1}} A_2$, the vector representation of the received signal at the destination can be written as

$$y_2 = \left(\bar{C}_2^T \otimes I_{n_d}\right) h_{sd} - \left(A_2^T N_{\hat{H}_{sd,1}}^T \otimes I_{n_d}\right) \Delta h_{sd,1} + w_2. \tag{5.86}$$

The LMMSE estimate of h_{sd} at the destination in stage 2 and the associated error covariance matrix is given respectively by

$$\hat{h}_{sd,2} = R_{h_{sd}y_2} R_{y_2 y_2}^{-1} y_2 \tag{5.87}$$

and

$$E[\Delta h_{sd,2} \Delta h_{sd,2}^H] = \sigma_{h_{sd}}^2 I_{n_d \times n_s} - R_{h_{sd}y_2} R_{y_2 y_2}^{-1} R_{h_{sd}y_2}^H, \tag{5.88}$$

where

$$R_{h_{sd}y_2} \triangleq E\left[h_{sd}y_2^H\right] = \sigma_{h_{sd}}^2 \left(\bar{C}_2^* \otimes I_{n_d}\right) \tag{5.89}$$

and

$$R_{y_2 y_2} \triangleq E\left[y_2 y_2^H\right] = \sigma_{h_{sd}}^2 \left(\bar{C}_2^T \bar{C}_2^* \otimes I_{n_d}\right)$$
$$+ E\left[\left(A_2^T N_{\hat{H}_{sd,1}}^T \otimes I_{n_d}\right) \Delta h_{sd,1} \Delta h_{sd,1}^H \left(A_2^T N_{\hat{H}_{sd,1}}^T \otimes I_{n_d}\right)^H\right] + \sigma_w^2 \left(I_{T_2} \otimes I_{n_d}\right). \tag{5.90}$$

In the nonreciprocal case, the LMMSE estimate and its corresponding NMSE cannot be evaluated in closed form due to the expectation in the second term of (5.90). This expectation involves the average over the channel estimates $\hat{H}_{ds,0}$ and $\hat{H}_{sd,1}$ obtained at the source in stage 0 and stage 1. To obtain a closed-form expression of the NMSE, we consider the following simplifying assumptions [4]:

(i) given $\hat{\mathbf{H}}_{ds,0}$, the LMMSE estimate $\hat{\mathbf{H}}_{sd,1}$ and the associated error matrix $\Delta\mathbf{H}_{sd,1}$ are statistically independent and (ii) the number of antennas at the source and the destination are sufficiently large, i.e, $n_s, n_d \gg 1$. By the above assumptions, the correlation matrix in (5.90) can be approximated as [4]

$$
\mathbf{R}_{\mathbf{y}_2\mathbf{y}_2}
$$

$$
\approx \left\{ \sigma_{h_{sd}}^2 \bar{\mathbf{C}}_2^T \bar{\mathbf{C}}_2^* + \left[(n_s - n_d)\sigma_{a,2}^2 \left(\sigma_{h_{sd}}^2 - \frac{\sigma_{h_{sd}}^4 P_{s1} T_1}{\sigma_{h_{sd}}^2 P_{s1} T_1 + n_s \sigma_w^2} \frac{n_s \mu}{\beta + n_s \mu} \right) + \sigma_w^2 \right] \mathbf{I}_{T_2} \right\} \otimes \mathbf{I}_{n_d},
$$
$$(5.91)$$

where $\mu \triangleq \dfrac{\sigma_{h_{ds}}^4 P_0 T_0}{\sigma_{h_{ds}}^2 P_0 T_0 + n_d \sigma_u^2}$. Using the above approximation and by choosing $T_2 = n_s$ and $\mathbf{C}_2$ such that $\mathbf{C}_2^H \mathbf{C}_2 = \mathbf{C}_2 \mathbf{C}_2^H = \mathbf{I}_{n_s}$, the NMSE at the destination can be approximated as

$$
\mathrm{NMSE}_d^{(2)} = \frac{\mathrm{tr}\left(\mathrm{E}\left[\Delta\mathbf{h}_{sd,2} \Delta\mathbf{h}_{sd,2}^H \right] \right)}{n_s n_d}
$$
$$(5.92)$$

$$
\approx \left[\frac{1}{\sigma_{h_{sd}}^2} + \frac{P_2 T_2/n_s}{(n_s - n_d)\sigma_{a,2}^2 \left(\sigma_{h_{sd}}^2 - \frac{\sigma_{h_{sd}}^4 P_{s1} T_1}{\sigma_{h_{sd}}^2 P_{s1} T_1 + n_s \sigma_w^2} \frac{n_s \mu}{\beta + n_s \mu} \right) + \sigma_w^2} \right]^{-1}.
$$
$$(5.93)$$

The NMSE at the eavesdropper can be computed as in the reciprocal case and is given by

$$
\mathrm{NMSE}_e^{(2)} = \left(\frac{1}{\sigma_{h_{se}}^2} + \frac{P_2 T_2/n_s}{(n_s - n_d)\sigma_{a,2}^2 \sigma_{h_{se}}^2 + \sigma_v^2} \right)^{-1}.
$$
$$(5.94)$$

Given the NMSE expressions in (5.92) and (5.94), the power allocation between training and AN can then be determined by minimizing the NMSE at the destination subject to a lower constraint on the NMSE at the eavesdropper. By letting $\mathcal{E}_0 \triangleq P_0 T_0$, $\mathcal{E}_{s1} \triangleq P_{s1} T_1$, $\mathcal{E}_{d1} \triangleq P_{d1} T_1$, and $\mathcal{E}_2 \triangleq P_2 T_2$, the optimization problem can be formulated as:

$$
\min_{\mathcal{E}_0, \mathcal{E}_{s1}, \mathcal{E}_{d1}, \mathcal{E}_2, \sigma_{a,2}^2 \geq 0} \mathrm{NMSE}_d^{(2)}
$$
$$(5.95a)$$

$$
\text{subject to } \mathrm{NMSE}_e^{(2)} \geq \gamma,
$$
$$(5.95b)$$

$$
\mathcal{E}_0 + \mathcal{E}_{s1} + \mathcal{E}_{d1} + \mathcal{E}_2 + (n_s - n_d)\sigma_{a,2}^2 T_2 \leq \bar{\mathcal{E}}_{tot},
$$
$$(5.95c)$$

$$
\mathcal{E}_0 + \mathcal{E}_{d1} \leq \bar{\mathcal{E}}_d,
$$
$$(5.95d)$$

$$
\mathcal{E}_{s1} + \mathcal{E}_2 + (n_s - n_d)\sigma_{a,2}^2 T_2 \leq \bar{\mathcal{E}}_s,
$$
$$(5.95e)$$

The problem is nonconvex but can also be efficiently handled using the monomial approximation and the condensation method [7] as done in the previous cases.

Example 5.3 Let us consider the scenario as in Example 5.2. In the nonreciprocal case, the training lengths are chosen as $T_0 = 2$ and $T_1 = T_2 = 4$, and the total lengths of the training signals emitted by the source and the destination are $T_{s,tot}=T_1+T_2=8$ and $T_{d,tot} = T_0 + T_1 = 6$, respectively. Let $\bar{P}_{ave} \triangleq \bar{\mathcal{E}}_{tot}/(T_{s,tot} + T_{d,tot})$ be the overall average transmit power constraint. Moreover, let $\bar{P}_s \triangleq \bar{\mathcal{E}}_s/T_{s,tot} = 30\,dB$ and $\bar{P}_d \triangleq \bar{\mathcal{E}}_d/T_{d,tot} = 20\,dB$ (relative to the noise variance) be the average power constraints at the source and the destination, respectively. In Fig. 5.5a, the NMSE at the destination and the eavesdropper is shown for varying values of the overall average power constraint $\bar{P}_{ave}$ and for two different NMSE constraints at the eavesdropper, i.e., $\gamma = 0.1$ and 0.03. Similar to that observed in Example 5.2, the NMSE at the destination decreases as $\bar{P}_{ave}$ increases while the NMSE at the eavesdropper is constrained above γ. In Fig. 5.5b, the corresponding pilot signal power values, i.e., P_0, P_{s1}, P_{d1}, and P_2, and the AN power $(n_s - n_d)\sigma_{a,2}^2$ are shown for varying values of $\bar{P}_{ave}$. We can also see that the pilot signal power utilized in stages 0 and 1 should be increased to allow for a better preliminary channel estimate and a more accurate AN placement.

5.3 Summary and Discussions

In this chapter, the DCE scheme was introduced as a technique to enhance physical layer secrecy through training and channel estimation. This approach is different from those introduced in previous chapters, which focus mainly on the signal design in the data transmission phase. Two DCE schemes were introduced, i.e., the feedback-and-retraining and the two-way training DCE schemes. In particular, a basic feedback-and-retraining DCE scheme consists of two stages, i.e., a preliminary training stage and a feedback-and-retraining stage. In the preliminary training stage, the source first emits a pure pilot signal to allow the destination to perform a rough estimate of the source-to-destination channel. In the feedback-and-retraining stage, the destination feeds back this rough estimate to the source, who then transmits an AN-assisted training signal to refine the destination's channel estimate. With the channel feedback information, AN can then be placed in the null space of the estimated channel to minimize its interference at the destination. The feedback-and-retraining process can be repeated multiple times to enhance the difference in channel estimation quality at the destination and the eavesdropper. This scheme is limited by the fact that the eavesdropper can also benefit from the preliminary training emitted by the source, and thus good performance discrimination cannot be achieved without multiple stages of feedback-and-retraining. In the two-way DCE scheme, the preliminary training signal is instead emitted by the destination to enable channel estimation directly at the source. When the channel is reciprocal, the source can

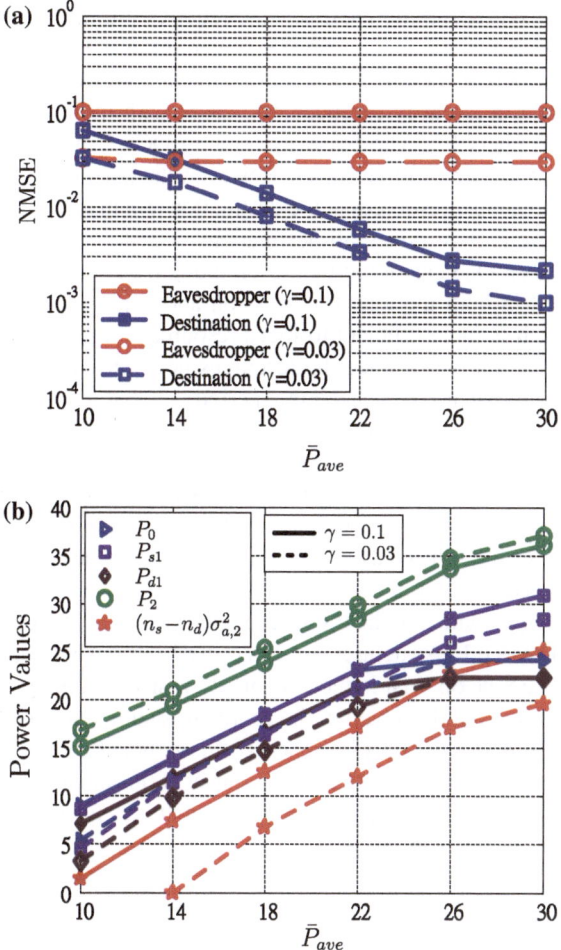

Fig. 5.5 NMSE performance and corresponding power allocation in the two-way DCE scheme for nonreciprocal channels. **a** NMSE performance, **b** power allocation between the pilot signal and AN powers

utilize its estimate of the backward channel to infer information about the forward channel and utilize the latter to determine the AN placement in the following forward training stage. When the channel is nonreciprocal, an additional round-trip training is required to provide the source with information of the combined forward and backward channels. Then together with the backward channel estimate obtained in the reverse training stage, the source would be able to compute an estimate of the forward channel. The AN-assisted forward training stage can then be performed similarly.

The feedback-and-retraining DCE scheme requires multiple stages of feedback and retraining, especially when the NMSE constraint at the eavesdropper is strict. This may be inefficient for fast fading environments. However, the advantage of this scheme is that it is easily extendable to cases with multiple destinations and multiple eavesdroppers [3]. The two-way training DCE scheme is more efficient since it requires only two or three stages. However, the performance of this scheme is reliant on the available resources at the destination and may be less efficient when multiple destinations exist (since all destinations need to perform the preliminary training separately). One can argue that, even though signal processing techniques can be employed in both the data transmission and channel estimation phases to enhance secrecy, it is more efficient to focus resources (e.g., for the use of AN) in the channel estimation phase since the resources need only be applied once every coherence interval (instead of each symbol period).

The original proposition of the DCE scheme (as described in [3, 4] and in this chapter) was derived using the LMMSE criterion. However, in practice, the eavesdropper may not be limited to using the LMMSE estimator. Hence, it would be interesting to investigate the performance of the DCE schemes using Cramer Rao Lower Bound (CRLB) as the measure of performance at the eavesdropper. Moreover, the DCE schemes described in this chapter utilize orthogonal or semi-unitary pilot matrices and often assume that the training lengths are equal to the number of transmit antennas. These pilot structures are often used in conventional point-to-point systems but may not be optimal in the case of AN-assisted training. Furthermore, in the schemes described in this chapter, the source is assumed to have more antennas than the destination and the eavesdropper. If this is not the case, it is possible to design the training sequence such that only a subset of the antennas at the destination are estimated at once. The optimal design of pilot matrices for DCE is an interesting direction for future studies in this area.

References

1. Hassibi B, Hochwald B (2003) How much training is needed in multiple-antenna wireless links? IEEE Trans Inf Theory 49(4):951–963
2. Yoo T, Goldsmith A (2006) Capacity and power allocation for fading MIMO channels with channel estimation error. IEEE Trans Inf Theory 52:2203–2214
3. Chang T-H, Chiang W-C, Hong Y-WP, Chi C-Y (2010) Training sequence design for discriminatory channel estimation in wireless MIMO systems. IEEE Trans Signal Processing 58(12):6223–6237
4. Huang C-W, Chang T-H, Zhou X, Hong Y-WP (2013) Two-way training for discriminatory channel estimation in wireless MIMO systems. IEEE Trans Signal Processing 61(10):2724–2738
5. Barhumi I, Leus G, Moonen M (2003) Optimal training design for MIMO OFDM systems in mobile wireless channels. IEEE Trans Signal Processing 51(6):1615–1624
6. Kay SM (1993) Fundamentals of statistical signal processing: estimation theory. Prentice Hall, Upper Saddle River
7. Boyd S, Kim S-J, Vandenberghe L, Hassibi A (2007) A tutorial on geometric programming. Optim Eng 8:67–127

8. Stüber GL (2011) Principles of mobile communication. Springer, Berlin
9. Zhou X, Lamahewa T, Sadeghi P, Durrani S (2010) Two-way training: optimal power allocation for pilot and data transmission. IEEE Trans Wireless Commun 9(2):564–569
10. Horn RA, Johnson CR (1991) Topics in matrix analysis. Cambridge University Press, Cambridge

Chapter 6
Enhancing Physical-Layer Secrecy in Modern Wireless Communication Systems

Abstract This chapter provides a brief survey of recent developments and applications of physical-layer secrecy techniques in modern wireless systems, such as cognitive radios, orthogonal frequency-division multiplexing (OFDM) systems, wireless ad hoc and multihop networks, and cellular networks. Some references to recent works in these areas are given and several challenging open problems are mentioned as potential research directions.

Keywords Secrecy · Cognitive radio · Orthogonal frequency-division multiplexing (OFDM) · Ad hoc and multihop networks · Cellular networks

6.1 Secrecy in Cognitive Radio Networks

Cognitive radio (CR) networks refer to systems of users that are able to sense the availability of the spectrum, to learn from the environment, and to adapt their operations based on these information [1]. Empowered by these capabilities, unlicensed or secondary users can communicate in the same spectrum as the primary user without causing excess interference to the latter. This can be done by having secondary users transmit opportunistically in identified spectrum holes, or adapt their transmissions to assist communication between primary users or to minimize their interference at the primary receivers. However, the existence of secondary users may affect the secrecy transmission between primary users and vice versa. Hence, extra measures must be taken to maintain secrecy in both primary and secondary systems.

Two physical-layer secrecy problems were considered in the literature on CR networks: (i) the problem of enhancing secrecy between primary users in the presence of secondary users' transmissions and (ii) the problem of achieving secrecy among secondary users subject to constraints set by the primary users. In the first case, secondary transmitters can structure their signals so that their interference may be used to disrupt the reception at the eavesdropper (c.f. Fig. 6.1) or act as relays to

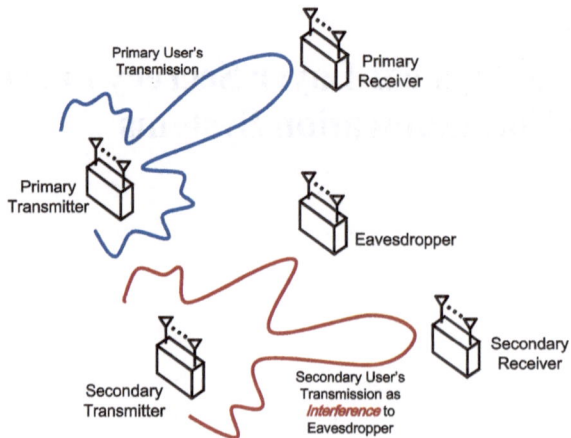

Fig. 6.1 Illustration of physical-layer secrecy enhancement between primary users by utilizing secondary users' transmissions as friendly interference at the eavesdropper

enhance the signal reception at the primary receiver. In the second case, secrecy among secondary users must be achieved while satisfying quality-of-service (QoS) constraints at the primary users or while preventing eavesdropping by primary users.

More specifically, in the first problem setting, confidential messages are to be communicated between primary transceivers and secrecy is to be maintained against a passive eavesdropper. These issues were studied recently in, e.g., [2, 3]. Without knowledge of the primary transmitter's message, the secondary transmitter can coordinate its transmissions with the primary users and act as friendly interferers to mask the transmission of confidential messages by the primary transmitter, as depicted in Fig. 6.1. This scenario constitutes an interference channel in the presence of an eavesdropper, but the goals of the primary and secondary users are asymmetric [4]. In particular, the primary transmitter's goal is to maximize the achievable secrecy rate between itself and the primary receiver, whereas the secondary transmitter's goal is to maximize the non-secrecy rate toward the secondary receiver. With accurate knowledge of the CSI, the primary and secondary users can negotiate their transmit powers to reach an optimal equilibrium between the secrecy rate in the primary system and the conventional information rate in the secondary system [2]. Moreover, if the secondary transmitter is able to decode the primary transmitter's message, it can also help relay the confidential message and, in this way, further improve the secrecy performance among primary users [3]. It is interesting to point out that, in traditional CR systems, primary users usually do not benefit from sharing their spectrum with secondary users and may in fact experience performance degradations due to collision or excess interference. However, as pointed out in [2], spectrum sharing may actually be beneficial to the primary users if confidential messages are to be transmitted.

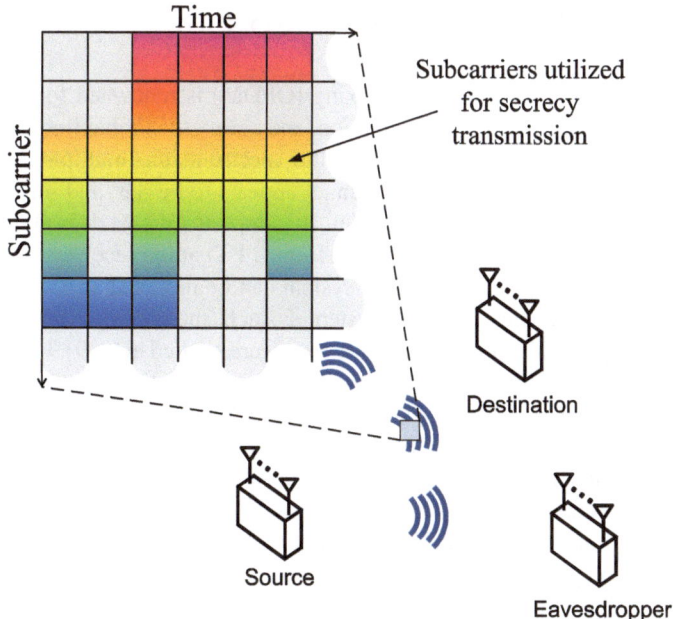

Fig. 6.2 Illustration of subcarrier assignment for OFDM systems with secrecy considerations

In the second problem setting, confidential messages are instead communicated between secondary transceivers, but QoS and/or secrecy constraints must be satisfied at the primary receiver and the eavesdropper, respectively. These issues were studied recently in, e.g., [5–7]. Specifically, this problem was examined in [5] from an information-theoretic perspective, where the secrecy capacity region of the so-called cognitive interference channel with confidential messages was derived. In this problem, the primary transmitter transmits its own message while the secondary transmitter helps transmit both its own and the primary user's message. The primary user's message is to be decoded by both primary and secondary receivers whereas the secondary user's message must be kept secret against the primary receiver (i.e., the eavesdropper in this case). In [6, 7], the use of multiple antennas at the secondary transmitter was considered. By assuming the existence of a separate eavesdropper, secrecy beamforming schemes were derived to maximize the secrecy rate at the secondary receiver, subject to an interference power constraint at the primary receiver and a secrecy constraint at the eavesdropper. The secrecy beamforming schemes proposed can be viewed as extensions to the techniques described in Chap. 3. The use of artificial noise to further enhance secrecy among secondary users is also an interesting research direction.

6.2 Secrecy in OFDM and OFDMA Systems

Orthogonal frequency-division multiplexing (OFDM) is renowned for its efficient spectrum utilization, simple implementation, and ability to combat frequency selective fading. In particular, OFDM divides the spectrum into multiple subcarriers, where the frequency response of the channel appears to be flat, and information is transmitted separately over each subcarrier. By adopting OFDM demodulators with cyclic prefix removal and fast Fourier transform (FFT) at the receivers, the channel input and output relations of an OFDM system can be modeled as a set of parallel Gaussian channels. By modeling the system as such, the secrecy capacity and the corresponding power allocation over subcarriers were studied in [8, 9]. In these cases, only the subcarriers that yield a better channel to the destination than the eavesdropper are utilized for secrecy transmissions, as depicted in Fig. 6.2. However, assuming that the eavesdropper adopts the OFDM receiver structure limits the eavesdropper's capability and, thus, may overestimate the achievable secrecy rate. In [10, 11], this assumption was relaxed and the channel input and output relations were modeled as a more general multiple-input multiple-output (MIMO) Gaussian wiretap channel. In this case, secrecy precoding can be performed over different subcarriers to either strengthen the signal at the destination or null out the signal toward the eavesdropper if the channel is known. The secrecy capacity reduction that is experienced when no constraint is placed on the eavesdropper's receiver structure was derived in [11].

Moreover, orthogonal frequency-division multiple access (OFDMA) has evolved as a leading technology for next generation multiuser high-speed wireless communication networks, such as the Third-Generation Partnership Project (3GPP) Long-Term Evolution (LTE), IEEE 802.16 worldwide interoperability for microwave access (WiMAX), and IEEE 802.22 wireless regional area networks (WRAN). In this system, a large variety of data traffic and QoS requirements can be efficiently addressed by performing power, subcarrier, and/or bit allocation [12]. In [13], power and subcarrier allocation policies were derived to maximize the sum secrecy capacity of a two-user secrecy broadcast channel. In [14], this problem was generalized to systems that consist of both secure and normal users, namely, users that require secure data transmission and users that demand only conventional best effort data, respectively. The optimal power and subcarrier allocation policies were found by maximizing the aggregate information rate of all normal users while maintaining a target average secrecy rate for each secondary user. A similar problem was examined in [15] from the point of view of achieving high energy efficiency. The study on physical layer secrecy in OFDM systems can also be extended to cases with relays, as done in [16, 17]. In this case, the relay strategy on each subcarrier must be chosen in addition to the power, subcarrier, and bit allocation. The use of friendly jamming by relays was also considered in these works.

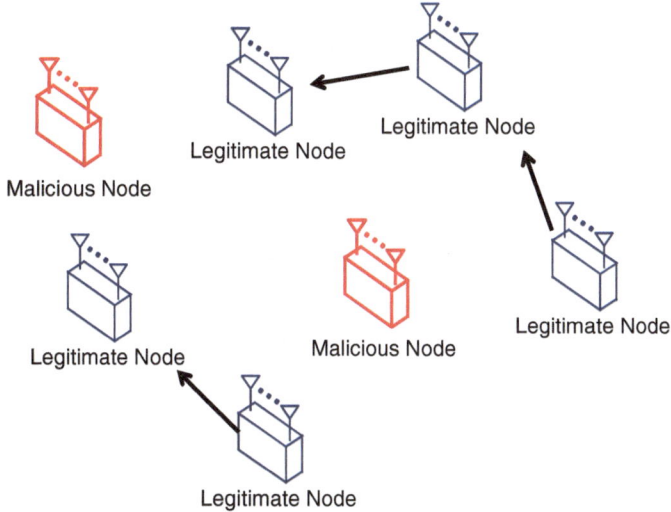

Fig. 6.3 Illustration of the ad hoc network with passive eavesdroppers

6.3 Secrecy in Ad Hoc and Multihop Networks

Wireless ad hoc networks consist of collections of wireless devices that are interconnected without a fixed infrastructure or pre-established communication links [18–20]. Nodes that are outside of the direct transmission range of each other can communicate through multihop connections, i.e., connections formed by the relaying of multiple intermediate nodes. These systems are desirable due to its ease of deployment, high robustness, distributed nature, etc. However, the use of multihop transmissions increases the system's vulnerability to eavesdropping and the lack of infrastructure makes it difficult for secret key distribution and management. Therefore, physical layer secrecy serves as a promising option for secret message or secret key sharing in ad hoc networks.

For multihop transmissions, the determination of a secure transmission path is critical. In [21], the authors utilized the concept of cumulative security leakage to study the impact of eavesdroppers in wireless multihop networks. A routing algorithm was proposed taking into consideration the cumulative security leakage on each route. Moreover, in [22], the impact of physical layer secrecy requirements on multihop transmissions was studied in the context of tree formation, where multiple legitimate nodes seek to obtain the most secure path toward a common destination in the presence of eavesdroppers. The interaction among legitimate nodes was modeled by a tree formation game in which each node decides its preferred path to the base station by optimizing a secrecy metric which reflects the level of security of the chosen path. Two choices of secrecy metrics were considered depending on whether the eavesdroppers' CSI is available, namely, the bottleneck secrecy rate of the path when CSI is available and the path qualification probability (i.e., the probability of

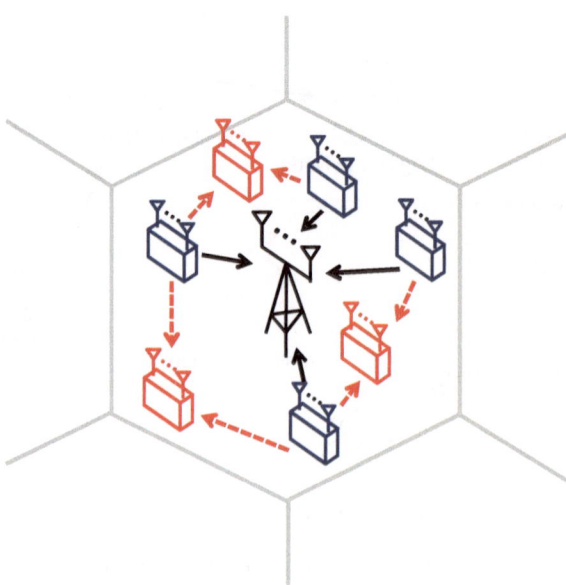

Fig. 6.4 Illustration of the cellular network with passive eavesdroppers

achieving a certain target secrecy rate over the path) when CSI is not available. A distributed algorithm that allows the nodes to interact and decide their paths was proposed in [22] based on the game-theoretic formulation.

Moreover, the impact of secrecy requirements on the performance of large stochastic networks has been studied recently in [23–26]. In particular, by using tools from stochastic geometry and percolation theory, it was shown in [23] that even a small density of eavesdroppers has a significant impact on the network connectivity. Here, secrecy is maintained as long as the eavesdroppers are not able to decode the message. Note that this requirement is less stringent than the secrecy constraint considered in typical wiretap channels, i.e., the constraint on the eavesdropper's equivocation rate. In [25, 26], a stronger notion of secrecy was considered in studies on the so-called intrinsically secure communications graph (iS-graph), i.e., a random graph that describes the connections that can be securely established over a large-scale network. The local connectivity of the Poisson iS-graph was established in terms of node degrees and isolation probabilities, and the achievable secrecy rates were derived. In the above, secrecy in large networks was studied with precise knowledge of the eavesdroppers channel gains. These issues were examined in the presence of eavesdropper uncertainties in [24].

Furthermore, the secrecy capacity scaling law has also been derived recently in [27, 28] for large ad hoc networks consisting of multiple legitimate and malicious nodes, as shown in Fig. 6.3. In particular, two cases were considered in [27]: The case where the malicious nodes act as passive eavesdroppers that only attempt to intercept the transmitted message and the case where the malicious nodes actively modify and deliver false messages to the destination. It was shown in [27, 28] that no loss in the capacity scaling law will be experienced if the number of eavesdroppers is below a certain critical threshold, which can be derived. This critical threshold is shown to be lower when the malicious nodes are active.

6.4 Secrecy in Cellular Networks

In cellular networks, secrecy in the downlink and uplink transmissions can be modeled as conventional broadcast and multiple access wiretap channels, respectively, in the physical layer. Coding and signal processing techniques, such as secrecy precoding and artificial noise usage, can then be developed to enhance secrecy as discussed in previous chapters. However, from a network perspective, it is also important to design network level protocols, such as scheduling and flow control, that can support or exploit properties of the secrecy transmission schemes in the physical layer. Specifically, in [29], the scheduling design with secrecy incorporated as a QoS metric is considered in a single-hop uplink cellular network, as depicted in Fig. 6.4. It is assumed that each node is a potential eavesdropper to other nodes within a broadcast channel with confidential messages. A dynamic joint flow control, scheduling, private encoding scheme is then proposed to optimize the achievable common and confidential message rates. In [30], a multicell network is considered, where two base stations are connected with a finite-capacity backbone. With the goal of securing the uplink transmission, one of the base stations is allowed to transmit the downlink data as interference to mask the secrecy transmission in the uplink. The receiving base station will be able to cancel the interference from the interfering base station with cooperation through the backbone.

References

1. Haykin S (2005) Govnitive radio: brain-empowered wireless communications. IEEE J Sel Areas Commun 23(2):201–213
2. Wu Y, Liu KJ (2011) An information secrecy game in cognitive radio network. IEEE Trans Inf Forensics Secur 6(3):831–842
3. Gabry F, Schrammar N, Girnyk M, Li N, Thobaben R, Rasmussen LK (2012) Cooperation for secure broadcasting in cognitive radio networks. In: Proceedings of IEEE International Conference on Communications, June 2012, pp 5613–5618
4. Tang X, Liu R, Spasojevic P, Poor HV (2011) Interference assisted secret communication. IEEE Trans Inf Theor 57(5):3153–3167
5. Liang Y, Somekh-Baruch A, Poor HV, Shamai S, Verdu S (2009) Capacity of cognitive interference channels with and without secrecy. IEEE Trans Inf Theory 55(2):604–619
6. Pei Y, Liang Y-C, Zhang L, Teh KC, Li KH (2010) Secure communication over MISO cognitive radio channels. IEEE Trans Wireless Commun 9(4):1494–1502
7. Pei Y, Liang Y-C, Teh KC, Li KH (2011) Secure communication in multiantenna cognitive radio networks with imperfect channel state information. IEEE Trans Signal Process 59(4):1683–1693

8. Li Z, Yates R, Trappe W (2006) Secrecy capacity of independent parallel channels. In: Proceedings of the 44th annual Allerton Conference on Communication, Control, and Computing

9. Rodrigues MRD, Almeida PDM (2008) Filter design with secrecy constraints: the degraded parallel Gaussian wiretap channel. In: Proceeings of IEEE Global Communications Conference (GLOBECOM), December 2008

10. Kobayashi M, Debbah M, Shamai S (2009) Secured communication over frequency-selective fading channels: A practical Vandermonde precoding. EURASIP J Wireless Commun Netw 2009:1–19. In (2012), 2009(4):1354–1367

11. Renna F, Laurenti N, Poor HV (2012) Physical-layer Secrecy for OFDM transmissions over fading channels. IEEE Trans Inf Forensics Secur 7(4):1354–1367

12. Wong CY, Cheng RS, Lataief KB, Murch RD (1999) Multiuser OFDM with adaptive subcarrier, bit, and power allocation. IEEE J Sel Areas Commun 17(10):1747–1758

13. Jorswieck EA, Wolf A (2008) Resource allocation for the wire-tap multi-carrier broadcast channel. In: Proceedings of International Conference on Telecommunications (ICT)

14. Wang X, Tao M, Mo J, Xu Y (2011) Power and subcarrier allocation for physical-layer security in OFDMA-based broadband wireless networks. IEEE Trans Inf Forensics Secur 6(3):693–702

15. Ng DWK, Lo ES, Schober R (2012) Energy-efficient resource allocation for secure OFDMA systems. IEEE Trans Veh Technol 61(6):2572–2585

16. Jeong C, Kim I-M (2011) Optimal power allocation for secure multicarrier relay systems. IEEE Trans Signal Process 59(11):5428–5442

17. Ng DWK, Lo ES, Schober R (2011) Secure resource allocation and scheduling for OFDMA decode-and-forward relay networks. IEEE Trans Wireless Commun 10(10):3528–3540

18. Royer EM, Toh C-K (1999) A review of current routing protocols for ad hoc mobile wireless networks. IEEE Pers Commun 6(2):46–55

19. Goldsmith AJ, Wicker SB (2002) Design challenges for energy-constrained ad hoc wireless networks. IEEE Wirel Commun 9(4):8–27

20. Ramanathan R, Redi J (2002) A brief overview of ad hoc networks: challenges and directions. IEEE Commun Mag 40(5):20–22

21. Bashar S, Ding Z (2009) Optimum routing protection against cumulative eavesdropping in multihop wireless networks. In: Proceedings of the IEEE Military Communications Conference (MILCOM)

22. Saad W, Zhou X, Maham B, Basar T, Poor HV (2012) Tree formation with physical layer security considerations in wireless multi-hop networks. IEEE Trans Wireless Commun 11(10):3980–3991

23. Haenggi M (2008) The secrecy graph and some of its properties. In: Proceedings of the IEEE International Symposium on Information Theory (ISIT), pp 539–543

24. Goel S, Aggarwal V, Yener A, Calderbank AR (2011) The effect of eavesdroppers on network connectivity: a secrecy graph approach. IEEE Trans Inf Forensics Secur 6(3):712–724

25. Pinto PC, Barros J, Win MZ (2012) Secure communication in stochastic wireless networks-part I: connectivity. IEEE Trans Inf Forensics Secur 7(1):125–138

26. Pinto PC, Barros J, Win MZ (2012) Secure communication in stochastic wireless networks-part II: maximum rate and collusion. IEEE Trans Inf Forensics Secur 7(1):139–147

27. Liang Y, Poor HV, Ying L (2011) Secrecy throughput of MANETs under passive and active attacks. IEEE Trans Inf Theory 10(57):6692–6720

28. Koyluoglu OO, Koksal CE, Gamal HE (2012) On secrecy capacity scaling in wireless networks. IEEE Trans Inf Theory 58(5):3000–3015

29. Koksal CE, Ercetin O, Sarikaya Y (2013) Control of wireless networks with secrecy. IEEE/ACM Trans Netw 21(1):324–337

30. Popovski p (2009) Wireless secrecy in cellular systems with infrastructure-aided cooperation. IEEE Trans Inf Forensics Secur 4(2):242–256

Index